Zongmin Ma

Fuzzy Database Modeling of Imprecise and Uncertain Engineering Information

T0135097

Studies in Fuzziness and Soft Computing, Volume 195

Editor-in-chief
Prof. Janusz Kacprzyk
Systems Research Institute
Polish Academy of Sciences
ul. Newelska 6
01-447 Warsaw
Poland
E-mail: kacprzyk@ibspan.waw.pl

Zongmin Ma

Fuzzy Database Modeling of Imprecise and Uncertain Engineering Information

 Springer

Dr. Zongmin Ma
College of Information Science & Engineering
Northeastern University
Shenyang, Liaoning 110004
People's Republic of China
E-mail: mazongmin@ise.neu.edu.cn

ISSN print edition: 1434-9922

ISBN   978-3-642-06795-2   e-ISBN   978-3-540-33013-4

Springer is a part of Springer Science+Business Media
springer.com
© Springer-Verlag Berlin Heidelberg 2006
Softcover reprint of the hardcover 1st edition 2006

Dedicated to My Parents and My Wife Li, My Daughter Ruizhe, and My Son Jiaji

Preface

Computer-based information technologies have been extensively used to help industries manage their processes and information systems hereby become their nervous center. More specially, databases are designed to support the data storage, processing, and retrieval activities related to data management in information systems. Database management systems provide efficient task support and database systems are the key to implementing industrial data management. Industrial data management requires database technique support. Industrial applications, however, are typically data and knowledge intensive applications and have some unique characteristics that makes their management difficult. Besides, some new techniques such as Web, artificial intelligence, and etc. have been introduced into industrial applications. These unique characteristics and usage of new technologies have put many potential requirements on industrial data management, which challenge today's database systems and promote their evolvement. Viewed from database technology, information modeling in databases can be identified at two levels: (conceptual) data modeling and (logical) database modeling. This results in conceptual (semantic) data model and logical database model. Generally a conceptual data model is designed and then the designed conceptual data model will be transformed into a chosen logical database schema. Database systems based on logical database model are used to build information systems for data management. Much attention has been directed at conceptual data modeling of industrial information systems. Product data models, for example, can be views as a class of semantic data models (i.e., conceptual data models) that take into account the needs of engineering data.

As it may be known, in many real-world applications, information is often vague or ambiguous. Therefore, different kinds of imperfect information have extensively been introduced and studied to model the real world as accurately as possible. In addition to complex structures and rich semantic relationships, one also needs to model imprecise and uncertain information in many industrial activities. Information imprecision and uncertainty exist in almost all engineering applications and has been investigated in the context of various engineering actions. Classical database models often suffer from their incapability of representing and manipulating imprecise and uncertain information. Since the early 1980's, Zadeh's fuzzy logic has been used to extend various database models in order to enhance the classical models such that uncertain and imprecise information can be represented and manipulated. This resulted in numerous contributions, mainly with respect to the popular relational model or to some related form of it. Since classical relational database model and its extension of fuzziness do not satisfy the need of modeling complex objects with imprecision and uncertainty, currently many researches have been concentrated on fuzzy conceptual data models and fuzzy object-oriented database models in order to deal with complex objects and uncertain data together. The research on fuzzy conceptual models and fuzzy object-oriented databases is receiving increasing attention in addition to fuzzy relational database model. It should be noticed, however, that there have been few efforts at investigating the issues of database modeling of imprecise and uncertain industrial information although databases have been widely applied in industrial applications.

The material in this book is the outgrowth of research the author has conducted in recent years. The topics include fuzzy conceptual data modeling of industrial information and database implementations of fuzzy conceptual data models for industrial information. Concerning the fuzzy conceptual data modeling of industrial information, in addition to the ER/EER and UML data models, the IDEF1X data model and the EXPRESS data model are extended for fuzzy industrial data modeling. Concerning the database implementations of fuzzy conceptual data models for industrial information, the conversion of the fuzzy conceptual data models to the fuzzy logical databases are investigated, in which the fuzzy logical databases include the fuzzy relational databases, fuzzy nested relational databases and fuzzy object-oriented databases. The mappings from the fuzzy IDEF1X model to the fuzzy relational databases and from the fuzzy EXPRESS-G model to the fuzzy nested relational databases are developed in addition to the mappings from the fuzzy ER model to the fuzzy relational databases and from the EER model to the fuzzy object-oriented databases. In particular, the object-oriented database implementation of the fuzzy EXPRESS model is introduced in this book.

This book aims to provide a single record of current research and practical applications in the fuzzy database modeling of industrial information. The objective of the book is to provide state of the art information to the researchers of industrial database modeling and while at the same time serve the information technology professional faced with a non-traditional industrial application that defeats conventional approaches. Researchers, graduate students, and information technology professionals interested in industrial databases and soft computing will find this book a starting point and a reference for their study, research and development.

I would like to acknowledge all of the researchers in the area of database modeling of industrial information and fuzzy databases. Based on both their publications and the many discussions with some of them, their influence on this book is profound. Much of the material presented in this book is a continuation of the initial research work that I did during my Ph.D. studies at City University of Hong Kong. I am grateful for the financial support from City University of Hong Kong through a research studentship. Additionally, the assistances and facilities of University of Saskatchewan and University of Sherbrooke, Canada, Oakland University and Wayne State University, USA, and Northeastern University and City University of Hong Kong, China, are deemed important, and are highly appreciated. Special thanks go to the publishing team at Springer-Verlag. In particular to series editors Dr. Janusz Kacprzyk and Dr. Thomas Ditzinger and to their assistant Heather King for their advice and help to propose, prepare and publish this book. This book will not be completed without the support from them. Finally I wish to thank my family for their patience, understanding, encouragement, and support when I needed to devote many time in development of this book. This book will not be completed without their love.

China   *Zongmin Ma*
October 2005

Contents

1 Engineering Information Modeling in Databases

1.1 Introduction

Information systems have become the nervous center of current computer-based engineering applications, which hereby put the requirements on engineering information modeling. Databases are designed to support data storage, processing, and retrieval activities related to data management, and database systems are the key to implementing engineering information modeling. Engineering information modeling requires database support. It should be noted that, however, the current mainstream databases are mainly used for business applications. Engineering applications are data and knowledge intensive applications. Some unique characteristics and usage of new technologies have put many potential requirements on engineering information modeling, which challenge today's database systems and promote their evolvement.

Database systems have gone through the development from hierarchical and network databases to relational databases. But in non-transaction processing such as CAD/CAPP/CAM (computer aided design/computer aided process planning/computer aided manufacturing), knowledge-based system, multimedia and Internet systems, most of these data intensive application systems suffer from the same limitations of relational databases. Therefore, some non-traditional data models have been proposed. These data models are the fundamental tools for modeling databases or the potential database models. Incorporation between additional semantics and data models has been a major goal for database research and development.

Information modeling, including engineering information modeling, in databases can be carried out at two different levels: *conceptual data modeling* and *logical database modeling*. Correspondingly, we have *conceptual data models* and *logical database models* for information modeling. Generally the conceptual data models are used for information modeling at a high level of abstraction and at the level of data manipulation (i.e., a low level of abstraction), the logical database model is used for information

Z. Ma: *Fuzzy Database Modeling of Imprecise and Uncertain Engineering Information*,
StudFuzz **195**, 1–32 (2006)
www.springerlink.com

modeling. Database modeling generally starts from the conceptual data models and then the developed conceptual data models are mapped into the logical database models. In the chapter, database models for engineering information modeling refer to conceptual data models and logical database models simultaneously.

1.2 Conceptual Data Models

Much attention has been directed at conceptual data modeling of engineering information (Mannisto *et al.*, 2001; McKay *et al.*, 1996). Product data models, for example, can be viewed as a class of semantic data models (i.e., conceptual data models) that take into account the needs of engineering data (Shaw *et al.*, 1989). Recently conceptual information modeling of enterprises such as virtual enterprises has received increasing attention (Zhang and Li, 1999). Generally speaking, traditional ER (entity-relationship) (Chen, 1976) and EER (extended entity-relationship) models can be used for engineering information modeling at conceptual level. But limited by their power in engineering modeling, some improved conceptual data models such as UML (Booch *et al.*, 1998), IDEF1X (IDEF, 2000; Kusiak *et al.*, 1997) and EXPRESS/STEP (ISO IS 10303-1 TC184/SC4, 1994; Schenck and Wilson, 1994) have been developed.

Table 1.1 gives some conceptual data models that may be applied for engineering information modeling.

Table 1.1. Conceptual data models for engineering information modeling

Conceptual Data Models	
Generic Conceptual Data Models	*Specific Conceptual Data Models for Engineering*
ER data model EER data model UML data model	IDEF1X data model EXPRESS data model

1.2.1 ER/EER Models

The *entity-relationship* (ER) model was incepted by P. P. Chen (Chen, 1976) and has played a crucial role in database design and information systems analysis. In spite of its wide applications, the ER model suffers from its incapability of modeling complex objects and semantic relationships.

So a number of new concepts have been introduced into the ER model by various researchers (dos Santos *et al.*, 1979; Elmasri *et al.*, 1985; Gegolla and Hohenstein, 1991; Scheuermann *et al.*, 1979) to enrich its usefulness and expressiveness, forming the notion of the *enhanced entity-relationship* (EER) model.

ER Model

The ER data model proposed by Chen (1976) can represent the real world semantics by using the notions of *entities, relationships*, and *attributes*. ER data schema described by the ER data model is generally represented by the *ER diagram*.

Entity. Entity is a concrete thing or an abstract notion that can be distinguishable and can be understood. A set of entities having the same characteristics is called an *entity set*. A named entity set can be viewed as the description of an entity type, while each entity in an entity set is an instance of that entity type. For example, car is an entity set. The descriptions of the features of a car belong to the entity type, while an actual modern car is an instance of the car entity.

Attribute and key. The characteristics of an entity are called *attributes* of the entity. Each attribute has a range of values, called a *value set*. Value sets are essentially the same as attribute domains in relational databases. The characteristics of an entity are called *attributes* of the entity. Each attribute has a range of values, called a *value set*. Value sets are essentially the same as attribute domains in relational databases. Like relational databases, a minimal set of attributes of an entity that can uniquely identify the entity is called a *key* of the entity. An entity may have more than one keys and one of them is designated as the *primary key*.

Relationship. Let $e_1, e_2, ..., e_n$ be entities. A relationship among them is represented as $r(e_1, e_2, ..., e_n)$. The relationship is 2-ary if n = 2 and is multiple-ary if n > 2. The set that consists of the same type of relationship is called *relationship set*. A relationship set can be viewed as a relationship among entity sets. $R(E_1, E_2, ..., E_n)$ denotes the relationship set defined on entity sets $E_1, E_2, ..., E_n$. Relationship set is the type description of the entity relationship and a relationship among concrete entities is an instance of the corresponding relationship set. The same entity set can appear in a relationship set several times. A named relationship set can be viewed as the description of a *relationship type*. Sometimes relationship type is called relationship for short. Note that relationships in the ER data model also have attributes, called the *relationship attributes*.

In the ER data model, a 2-ary relationship can be one-to-one, one-to-many, or many-to-many relationships. This classification can be applied to n-ary relationships as well. The constraint of a relationship among entities is called *cardinality ratio constraint*. In the ER data model there is an important semantic constraint called *participation constraint*, which stipulates the way that entities participate in the relationship. The concept *participation degree* is used to express the minimum number and maximum number of an entity participating in a relationship, expressed as (min, max) formally, where max $\geq$ min ≥ 0 and max ≥ 1. When min $= 0$, the way an entity participates in a relationship is called *partial participation*, and is called *total participation* otherwise. The cardinality ratio constraint and participation constraint are, sometimes, referred to as the structure constraint.

There is a special relationship in the real world, which represents the ownership among entities and is called the *identifying relationship*. Such a relationship has the characteristics as follows.

- The entity owned by another entity depends on an owning entity, and does not exist separately, which must totally participate in relationship.
- The entity owned by another entity may not be the entity key of itself.

Because the entity owned by another entity has such characteristics, it is called the *weak entity*. A weak entity can be regarded as an entity as well as a complex attribute of its owning entity.

ER diagram. In the ER diagram, entities, attributes and relationships should be represented, where a rectangle denotes an entity set, a rectangle with double lines denotes a weak entity set, and a diamond denotes a relationship. Rectangles and rhombus are linked by arcs and the cardinality ratios of relationships are given. If an arc is a single line, it represents that the entity is a partial participation. If an arc is a double line, it represents that the entity is a total participation. Participation degrees may be given if necessary.

In the ER diagram, a circle represents an attribute and it is linked to the corresponding entity set with an edge. If an attribute is an entity key or a part of the entity key, it is pointed out that in the ER diagram by underlining the attribute name or adding a short vertical line on the edge. If an attribute is complex, a tree structure will be formulated in the ER diagram.

EER Model

The ER model based on *entities*, *relationships* and *attributes* is called the basic ER model. In order to model the complex semantics and relationships in the applications such as CAD/CAM, CASE, GIS, and so on, some new concepts have been introduced in the ER model and the *enhanced (extended) entity-relationship* (EER) data model is formed. In the EER model, the following notions are introduced.

Specialization and generalization. Generalization can summarize several entity types with some common features to an entity type and define a superclass. Specialization can divide an entity type into several entity types according to a particular feature and define several subclasses. For example, entity type "Automobile" is specialized into several subclasses such as "Car" and "Truck" while entity types "Faculty", "Technician", and "Research Associate" are generalized into a superclass "Staff".

Symbolically, a superclass E and several subclasses $S_1, S_2, ..., S_n$ satisfy the relationship $S_1 \cup S_2 \cup ... \cup S_n \subseteq E$. Let $F = \cup_i S_i$ $(1 \leq i \leq n)$. Then if $F = E$, F is a total specialization of E, or it is a partial one. In addition, F is a disjoint if $S_i \cap S_j = \Phi$ $(i \neq j)$, or it is overlapping with $G = \cup_i S_i$ $(1 \leq i \leq n)$. It should be noted that a subclass may not only inherit all attributes and relationships of its superclasses, but also have itself attributes and relationships.

Category. A category is a subclass of the union of the superclasses with different entity types. For example, entity type "Account" may be entity types "Personal" or "Business". Symbolically, a category E and the supclasses $S_1, S_2, ..., S_n$ satisfy the relationship $E \subseteq S_1 \cup S_2 \cup ... \cup S_n$. The difference between the category and the subclass with more than one superclass should be noticed. Let E be a subclass and $S_1, S_2, ..., S_n$ be its superclasses. One has then $E \subseteq S_1 \cap S_2 \cap ... \cap S_n$.

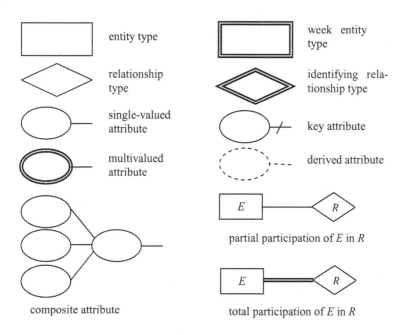

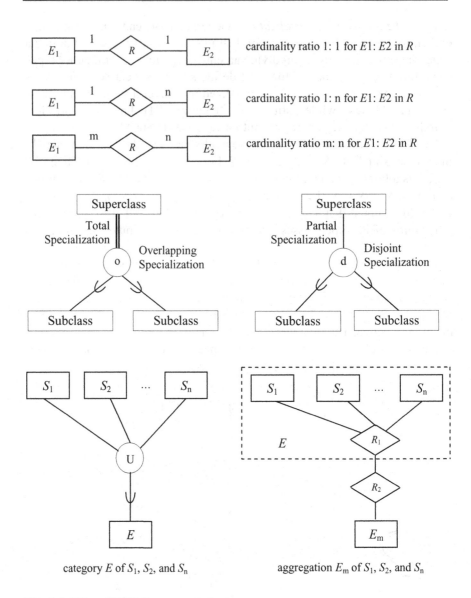

cardinality ratio 1: 1 for $E1$: $E2$ in R

cardinality ratio 1: n for $E1$: $E2$ in R

cardinality ratio m: n for $E1$: $E2$ in R

category E of S_1, S_2, and S_n

aggregation E_m of S_1, S_2, and S_n

Fig. 1.1. ER and EER diagram notations

Aggregation. A number of entity types, say S_1, S_2, ..., S_n, are aggregated to form an entity type, say E. In other words, E consists of S_1, S_2, ..., and S_n. For example, an entity type "Automobile" is aggregated from some entity types such as "Engine", "Gearbox", and "Interior", where "Interior" consists of "Seat" and "Dashboard". Here, S_i (i = 1, 2, ..., n) can be viewed as a kind of composite attribute, but it is not a simple attribute, which is an entity type consisting of simple attributes or other entity types. Therefore, the aggregation abstract is proposed in object-oriented modeling as an abstract means. Being not the same as specialization/generalization abstract, aggregated entity and all component entities belong to different entity types.

Figure 1.1 shows ER diagram notations and some new symbols introduced into the EER diagram.

1.2.2 UML Class Model

The Unified Modeling Language (UML) (Booch *et al.*, 1998; OMG, 2001) is a set of OO modeling notations that has been standardized by the Object Management Group (OMG). The power of the UML can be applied for many areas of software engineering and knowledge engineering (Mili *et al.*, 2001). The complete development of relational and object relational databases from business requirements can be described by the UML. The database itself traditionally has been described by notations called entity relationship (ER) diagrams, using graphic representation that is similar but not identical to that of the UML. Using the UML for database design has many advantages over the traditional ER notations (Naiburg, 2000). The UML is based largely upon the ER notations, and includes the ability to capture all information that is captured in a traditional data model. The additional compartment in the UML for methods or operations allows you to capture items like triggers, indexes, and the various types of constraints directly as part of the diagram. By modeling this, rather than using tagged values to store the information, it is now visible on the modeling surface, making it more easily communicated to everyone involved. So more and more, the UML is being applied to data modeling (Ambler, 2000a; Ambler, 2000b; Blaha and Premerlani, 1999; Naiburg, 2000). More recently, the UML has been used to model XML conceptually (Conrad *et al.*, 2000).

From the database modeling point of view, the most relevant model is the class model. The building blocks in this class model are those of classes and relationships. We briefly review these building blocks in the following.

Class

Being the descriptor for a set of objects with similar structure, behavior, and relationships, a class represents a concept within the system being modeled. Classes have data structure and behavior and relationships to other elements.

A class is drawn as a solid-outline rectangle with three compartments separated by horizontal lines. The top name compartment holds the class name and other general properties of the class (including stereotype); the middle list compartment holds a list of attributes; the bottom list compartment holds a list of operations. Either or both of the attribute and operation compartments may be suppressed. A separator line is not drawn for a missing compartment. If a compartment is suppressed, no inference can be drawn about the presence or absence of elements in it.

Relationships

Another main structural component in the class diagram of the UML is relationships for the representation of relationship between classes or class instances. UML supports a variety of relationships.

Aggregation and composition. An aggregation captures a whole-part relationship between an aggregate, a class that represent the whole, and a constituent part. An open diamond is used to denote an aggregate relationship. Here the class touched with the white diamond is the aggregate class, denoting the "whole".

Aggregation is a special case of composition where constituent parts directly dependent on the whole part and they cannot exist independently. Composition mainly applies to attribute composition. A composition relationship is represented by a black diamond.

Generalization. Generalization is used to define a relationship between classes to build taxonomy of classes: one class is a more general description of a set of other classes. The generalization relationship is depicted by a triangular arrowhead. This arrowhead points to the superclass. One or more lines proceed from the superclass of the arrowhead connecting it to the subclasses.

Association. Associations are relationships that describe connections among class instances. An association is a more general relationship than aggregation or generalization. A role may be assigned to each class taking part in an association, making the association a directed link. An association relationship is expressed by a line with an arrowhead drawn between the participating classes.

Dependency. A dependency indicates a semantic relationship between two classes. It relates the classes themselves and does not require a set of instances for its meaning. It indicates a situation in which a change to the target class may require a change to the source class in the dependency. A dependency is shown as a dashed arrow between two classes. The class at the tail of the arrow depends on the class at the arrowhead.

Figure 1.2 shows UML class diagram notations.

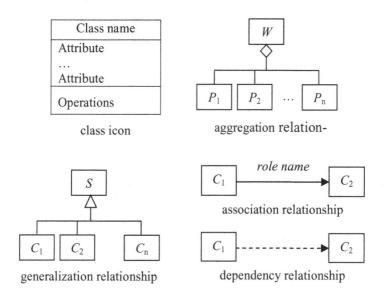

Fig. 1.2. UML class model notations

1.2.3 IDEF1X Model

Stemmed from the entity-relationship (ER) approach to semantic data modeling developed by Chen (1976), the IDEF1X methodology shares many of the same constructs proposed in entity-relationship models, namely, *entities*, *attributes*, and *relationships* between these entities (Kusiak *et al.*, 1997). Hence, IDEF1X is most useful for logical database design after the information requirements are known and the decision to implement a relational database has been made.

Entity

An entity is a class of real or abstract objects with the same characteristics. An individual member of the entity is called an entity instance. In IDEF1X data model, entities are classified to be identifier independent and identifier dependent. For an identifier independent entity, each of its instances can be uniquely identified without determining its relationship to another entity.

An entity is represented by a rectangle, where the box for identifier-independent entity has square corners whereas the box for identifier-dependent entity has rounded corners. In addition, each entity is assigned a unique name and number, which are placed above the box and separated with a '/'.

Attributes

The characteristics of an entity are called attributes of the entity, e.g. employee number, part number, part quantity, and machine capacity. An attribute takes a range value. For the Staff entity, for example, one attribute might be 'Age' and its range of taking value might be from 25 to 60. The value of the attribute 'Age' of the entity instance John might be 35.

Primary and alternative keys. A minimal set of attributes of an entity whose values can uniquely identify each entity instance is called the *key* of the entity. An entity may have more than one key and we designate one of them as the *primary key*, while other candidate keys are known as *alternative keys*. Assume that the Staff entity has attributes 'Staff number', 'Staff name', 'Social security number', 'Salary', and 'Birth date'. The 'Staff number' attribute and 'Social security number' attribute would then respectively be the primary key and the alternative key. Since it would be possible to have more than one staff with the same name or the same birth date, 'Staff name' attribute or 'Birth date' attribute cannot be a key. But the union of these two attributes can be the alternative key.

Foreign keys. In an entity, there may exist such attribute or attribute set that is the primary key of another entity. Such attribute or attribute set is known as the *foreign key* of the given entity. Not being the same as relational databases where foreign key can only be non-key attribute, in IDEF1X, however, the foreign key may be a primary key, a part of a primary key, an alternate key, or a non-key attribute of the given entity.

With the foreign key, the specific or categorization relationship between two entities can be established. At this moment, the foreign key in the child or category entity inherits the primary key of the parent or generic entity. If all the primary key attributes of a parent entity are inherited as part of the primary key of the child entity, then the relationship between the parent and child is an identifying relationship. If any of the inherited attributes are not part of the primary key, then it is a non-identifying relationship. In a categorization relationship, both the generic entity and the category entities represent the same thing. The primary key for all category entities is inherited through the categorization relationship from the primary key of the generic entity.

Constraint rules on the attributes of the entity. Some constraints on the attributes of entities can be identified in an IDEF1X model as follows (SofTech, 1981; WIZDOM, 1985; Kusiak *et al.*, 1997).

- The single-owner rule. It means that each attribute is used by only one entity.
- The no-null rule. It means that each instance of the entity must have a value for each attribute of the entity, i.e. the attribute must be applicable to every instance associated with the entity.
- The no-repeat rule. It means that no entity instance can have more than one value for any attribute of the entity.
- The smallest-key rule. It means that if K is a primary key, there is no such K' that $K' \subset K$ and K' is also the key.
- The full-functional-dependency rule. It means that if the primary key is composed of more than one attribute, the value of every non-key attribute must be functionally dependent on the entire primary key, i.e. no non-key attributes can be determined by just part of the primary key.
- The no-transitive-dependency rule. It means that each non-key attribute must be only functionally dependent on the primary and alternative keys, i.e. no non-key attribute's value can be determined by a non-key attribute value.
- The unified-reference rule. It means that the value of the foreign key must be identical to the value of the primary key of the parent or generic entity.

Attributes are shown by placing each attribute name on one line inside the entity box. Primary key attributes are placed at the top of the list, separated from other attributes by a line across the entity box. Alternative keys are identified by placing '(AKn)' next to the attribute, where *n* is a unique integer. A foreign key is identified by placing '(FK)' after the attribute. If the foreign key belongs to the primary key of the entity, it is placed above the horizontal line and the entity box will have rounded corners. If the foreign key does not belong to the entity's primary key, it is placed below the horizontal line and the box will be square.

Relationships

The relationships of IDEF1X represent the relationships between entities. Three kinds of relationships of IDEF1X can be identified, which are *connection relationship, categorization relationship* and *nonspecific relationship*.

Connection relationship. A connection relation is an association between two entities, among which one is called the parent entity and another is called the child entity. Here, each instance of the child entity is associated with exactly one instance of the parent entity whereas each instance of the parent entity is somehow related to zero, one, or more than one instances of the child entity. The number of child entity instances for each parent entity instance is defined as the *cardinality* of the connection relationship.

In an IDEF1X model, the following cardinality situations can be identified (Appleton 1986).

- Each parent entity instance may have zero, one, or more child entity instances.
- Each parent entity instance must have exactly zero or exactly one child entity instance.
- Each parent entity instance must have at least one or more child entity instances.
- Each parent entity instance has exactly some number of child entity instances.

If an instance of the child entity is identified by its association with the parent entity, then the relationship is referred to as an identifying connection relationship. Otherwise, it is a non-identifying connection relationship.

A connection relationship is represented by a line from the parent entity to a child entity, with a dot on the child end of the line and a relationship name is placed beside the line. The cardinality of the relationship should be indicated. A 'P' placed next to the dot signifies that it is a one or more relationship and a 'Z' next to the line signifies that it is a zero or one relationship. When the cardinality of the relationship is a specific number, then the positive integer corresponding to that cardinality is placed next to the dot. The default cardinality for such a relationship is zero, one, or more. An identifying relationship is represented with a solid line. It is clear that the child in an identifying relationship is always an identifier-dependent entity, which is represented as a box with rounded corners. Note that the parent entity in an identifying relationship would be an identifier-independent entity or an identifier-dependent entity, depending on if it is a child in any identifying relationship. A non-identifying relationship is represented with a dashed line. It is clear that in this situation, children and parent entities can be identified independently of one another. Both parent and child entities are identifier-independent entities in a non-identifying relationship, unless either is a child entity in an identifying relationship.

Sometimes a child will have more than one relationship with the same parent entity and these relationships are named differently. If they are identifying relationships, the child will inherit the parent's primary key more than once and each inherited attribute instance might have different values. When this occurs, a role name is assigned to each occurrence to distinguish them from each other. Role names are separated from the attribute names by a period. Let us look at two entities 'Part' and 'Assbl-struct'. Here, a particular part might be a component in a structure, and also the part might be assembled from another assembly structure. If the primary key of 'Part' is 'Part-num', then 'Part-num' would be inherited twice by 'Assbl-struct'. Role names Comp-num and Assbl-num are connected with Part-num as the primary key of the child entity 'Assbl-struct'. In addition, one may also define assertions about multiple relationships such as the exclusive-or logical operator, meaning that for a given parent entity instance, if one type of child entity instance exists, any other kind must not exist. This is a categorization relationship, which is described later.

Categorization relationship. A categorization relationship means that some entities, called category entities, are categories of other entities, called the generic entities. For example, Academic staff and Non-academic staff entities would be categories of the Employee entity. Note that the category entities for a generic entity must be mutually exclusive in IDEF1X. One person cannot be both an academic staff and a non-academic staff. Overlapping is not permitted.

Two kinds of categorization relationship can be identified in IDEF1X model.

- Complete categorization relationship. It means that each instance of the generic entity is associated with exactly one instance of the category entities.
- Incomplete categorization relationship. It means that a generic entity instance may not be contained in any instance of the category entities.

Since the generic and the category entities represent the same real-world object, they have the same primary key attributes. But it is clear that the primary key of the category entities is inherited from that of the generic entity, so the category entities are identifier-dependent. The generic entity is independent unless it is a child in an identifying relationship. In addition, there should be an attribute in the generic entity whose value can uniquely identify which of the categories each instance of the generic entity belongs to. This attribute is known as the discriminator. All instances of a category entity have the same discriminator value and all instances of different categories have different discriminator values. Note that a category entity might be a generic entity and a generic entity might be a category entity in other categorization relationships. That implies that categorizations may be nested. In addition, it is possible that a generic entity may have any number of categorization relationships. A category entity cannot be a child entity in an identifying relationship.

A categorization relationship is represented with a line from the generic entity to an underlined circle and lines from the circle to each of the category entities. The name of the discriminator is written next to the circle. For a complete categorization, the circle is double underlined. For an incomplete categorization, the circle is single underlined.

Nonspecific relationship. A non-specific relationship can be referred to a many-to-many relationship between two entities, where each entity instance in one relationship is associated with zero, one, or many entity instances in another relationship. For example, a course can be enrolled by many students and each student can enroll many courses. The relationship between the courses and the students is non-specific.

Being similar to the connection relationships, the cardinality at each end of the non-specific relationship may be zero, one, or more; one or more; zero or one; or an exact number. If a cardinality of exactly one exists at each end of the relationship, then the relationship is specific rather than non-specific. Therefore, connection and categorization relationships are considered to be 'specific relationships' because they show precisely how one entity instance relates another.

A non-specific relationship is represented with a line with a dot on each end drawn between two entities. Two relationship names are placed beside the line: the first name corresponds to the leftmost or topmost entity in the relationship and the second name corresponds to the rightmost or bottom entity, depending on the relationship's orientation. The notation '/' is used to separates these two names.

Figure 1.3 shows the syntaxes of IDEF1X.

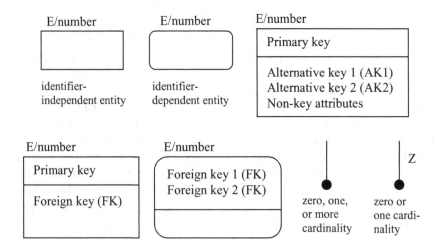

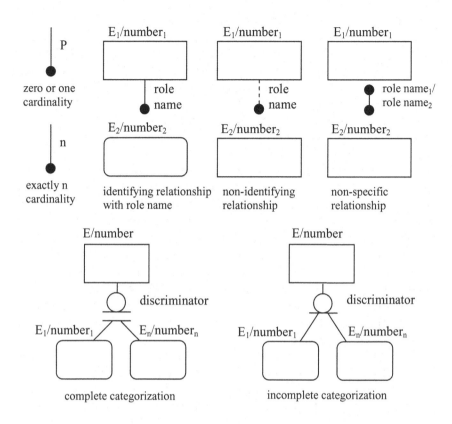

Fig. 1.3. IDEF1X syntax

1.2.4 EXPRESS Model

To share and exchange product data, the Standard for the Exchange of Product Model Data (STEP) is being developed by the International Organization for Standardization (ISO). STEP provides a means to describe a product model throughout its life cycle and to exchange data between different units. STEP consists of 4 major categories, namely, *description methods, implementation methods, conformance testing methodology and framework*, and *standardized application data models/schemata*. EXPRESS (Schenck and Wilson, 1994), as the description methods of STEP and a conceptual schema language, can model product design, manufacturing, and production data. EXPRESS model hereby becomes a one of the major conceptual data models for engineering information modeling.

The purpose of the information modeling process with EXPRESS, in general, centers on the descriptions of the objects that you create to represent information of interest to you. The main objects to be described include the *data types* and *declarations* as well as the *expressions, executable statements*, and *interfacing*. In addition, EXPRESS provides a graphical representation of EXPRESS, called *EXPRESS-G*, which uses graphical symbols to conceptually design an EXPRESS model and form a diagram. Note that EXPRESS-G can only represent a subset of the full language of EXPRESS.

Entity definitions. Entity definitions describe classes of real-world objects with associated properties. The properties are called attributes and can be simple values, such as "name" or "weight," or relationships between instances, such as "owner" or "part of". Attributes in EXPRESS are classified into explicit one for which the values are provided directly, derived one for which the values can be calculated from other attributes, and inverse one which captures the relationship between the entity being declared and a named attribute. Entities can also be organized into classification hierarchies, and inherit attributes from supertypes. The inheritance model supports single and multiple inheritance, as well as a new type, called AND/OR inheritance.

Type definitions. Type definitions describe ranges of possible values. The language provides several built-in types, and modeler can construct new types using the built-in types, generalizations of several types, and aggregates of values. Data types are used to represent domains of instance values. There are more rich data types in EXPRESS, six kinds of data types in particular.

- Simple types. These are the most basic data types, i.e., *number, integer, real, string, logical, Boolean*, and *binary*.
- Aggregate types. The collections of elements of some basic data types construct aggregate types, in which the elements may be ordered or may be unordered. *Array, bag, list and set* all belong to aggregate types.
- Entity type. This is an object type declared by entity declarations.
- Defined type. This is a type that is defined by the user and can be seen as an extension form to the set of standard data types.
- Enumeration type. This is an ordered set of values represented by names.
- Select or generic type. This type defines a named collection of other types. This collection is called the select list. It allows an attribute or a variable to be one of several possible types.

Correctness rules. A crucial component of entity and type definitions is local correctness rules. These local rules constrain relationships between entity instances or define the range of values allowed for a defined type. Global rules can also make statements about an entire information base. Local rules are assertions about the validity of entity instances, and apply to all instances of that entity type. The local rules are specified in the context of an entity declaration. There are two kinds of local rules as follows.

- Uniqueness rule or constraint. This controls the uniqueness of attribute values among all the instances of a given entity type.
- Domain rules or constraints.

Algorithmic definitions. An information modeler may also define functions and procedures to assist in the algorithmic description of constraints.

Relationships. There are classifications among entities in EXPRESS. The subtype/supertype structure is the mechanism for specifying classifications in which a subtype is a specialization of its supertype and a supertype is a generalization of a subtype. The subtypes of a supertype have several relationships among themselves.

- One-of relationship. This restricts the subtype instances to being mutually exclusive.
- And-or relationship. This defines that the subtypes are not mutually exclusive, and that instances may belong to more than one subtype.
- And relationship. This allows the definition of multiple, mutually exclusive relations as alternatives.

A subtype may inherit all or part of attributes from its one or multiple supertypes if there are. When a subtype inherits some attributes from more than one supertype, there may be name conflicts to be resolved. Besides the inheritance of attributes, subtype can also inherit rules from its supertype(s).

The information model in EXPRESS consists of some data schema. In order to reduce the redundancies among different schemas, two kinds of specifications *USE* and *REFERENCE* are used. The difference between USE and REFERENCE is that USE imports the external declaration to make it local and modifiable, while REFERENCE allows the access of an external declaration.

EXPRESS-G

EXPRESS-G is the graphical representation of EXPRESS, which uses graphical symbols to form a diagram (Eastman and Fereshetian, 1994). Note that it can only represent a subset of the full language of EXPRESS. EXPESS-G provides supports for the notions of entity, type, relationship, cardinality, and schema. The functions, procedures, and rules in EXPRESS language are not supported by EXPRESS-G.

EXPRESS-G has three basic kinds of symbol for definition, relationship, and composition. Definition and relation symbols are used to define the contents and structure of an information model. Composition symbols enable the diagrams to be spread across many physical pages.

Definition symbols. A definition symbol is a rectangle enclosing the name of the thing being defined. The type of the definition is denoted by the style of the box. Symbols are provided for simple type, defined type, entity type, and schema.

- Simple type symbols. A number of predefined simple types offered by EXPRESS language include Binary, Boolean, Integer, Logical, Number, Real, and String. The symbol for them is a solid rectangle with a double vertical line at its right end. The name of the type is enclosed within the box.
- Type symbols.
- Entity symbols.
- Schema symbols.

Relationship symbols. There are three different styles of lines, namely, a dashed line, a thick solid line, and a thin solid line, which are employed to connect related definition symbols. A relationship for an optional valued attribute of an entity is displayed as a dashed line, as is a schema-schema reference. A supertype-subtype relationship is displayed as a thick solid line. All other relationships are displayed as thin lines. Note that the lines with open circles denote relationship directions in EXPRESS-G.

Composition symbols. Graphical representation of models often spans many pages. Each page in a model must be numbered so that we can keep track of where we are in the model and enable inter-page referencing. In addition, a schema may utilize definitions from another schema. Therefore, there are two kinds of composition symbols for page references and inter-schema references. EXPRESS-G provides two levels of modeling, namely, *schema level model* and *entity level model.* Therefore we discuss the fuzziness in the entity level model and in the schema level model in the following, respectively.

Figure 1.4 shows EXPRESS-G diagram notations.

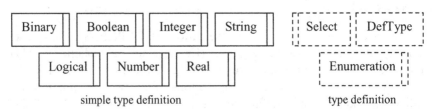

simple type definition   type definition

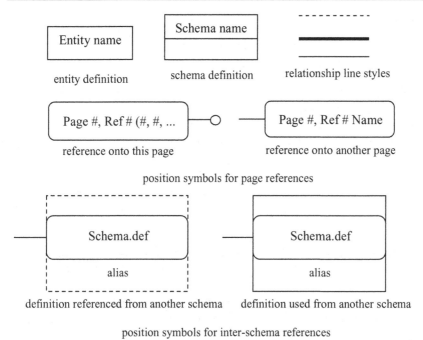

Fig. 1.4. EXPRESS-G diagram notations

1.3 Logical Database Models

As to engineering information modeling in database systems, the generic logical database models such relational databases, nested relational databases, and object-oriented databases can be used. Table 1.2 gives some logical database models that may be applied for engineering information modeling.

In the following, we only discuss the classical logical database models.

1.3.1 Relational Database Model

Relational database model introduced first by Codd (1970) is the most successful one and relational databases have been extensively applied in most information systems in spite of the increasing populations of object-oriented databases.

Table 1.2. Database models for engineering information modeling

Logical Database Models	
Classical Logical Database Models	*Specific & Hybrid Database Models*
Relational databases Nested relational databases Object-oriented databases Object-relational databases	Active databases Deductive databases Constraint databases Spatio-temporal databases Object-oriented active databases Deductive object-relational databases …

Attribute and domain. The representations for some features are usually extracted from real-world things. The features of a thing are called *attributes*. For each attribute, there exists a range that the attribute takes values, called *domain* of the attribute. A domain is a finite set of values and every value is an atomic data, the minimum data unit with meanings.

Relation and tuple. Let $A_1, A_2, ..., A_n$ be attribute names and the corresponding attribute domains be $D_1, D_2, ..., D_n$ (or Dom (A_i), $1 \leq i \leq n$), respectively. Then relational schema R is represented as

$$R = (D_1/A_1, D_2/A_2, ..., D_n/A_n) \text{ or } R = (A_1, A_2, ..., A_n),$$

where n is the number of attributes and is called the degree of relation. The instances of R, expressed as r or r (R), are a set of n-tuples and can be represented as $r = \{t_1, t_2, ..., t_m\}$. A tuple t can be expressed as $t = <v_1, v_2, ..., v_n>$, where $v_i \in D_i$ ($1 \leq i \leq n$), i.e., $t \in D_1 \times D_2 \times ... \times D_n$. The quantity r is therefore a subset of Cartesian product of attribute domains, i.e., $r \subseteq D_1 \times D_2 \times ... \times D_n$. Viewed from the content of a relation, a relation is a simple table, where tuples are its rows and attributes are its columns. Note that there is no complex data in relational table. The value of a tuple t on attribute set S is generally written $t [S]$, where $S \subseteq R$.

Keys. If an attribute value or the values of an attribute group in a relation can solely identify a tuple from other tuples, the attribute or attribute group is called a *super key* of the relation. If any proper subsets of a super key are not a super key, such super key is called a *candidate key* or shortly *key*. For a relation, there may be several candidate keys. One chooses one candidate as the *primary key,* and other candidates are called *alternate key.* It is clear that the values of primary key of all tuples in a relation are different and are not null. The attributes included in a candidate key are called *prime attributes* and not included in any candidate key called *nonprime attributes*. If an attribute or an attribute group is not a key of relation *r* but it is a key of relation s, such attribute (group) is called *foreign key* of relation *r*.

Constraints. There are various constraints in the relational databases. We identify the following integrity constraints.

- *Domain integrity constraints.* The basic contents of domain integrity constraints are that attribute values should be the values in the domains. In addition, domain integrity constraints are also prescribed if an attribute value could be null.
- *Entity integrity constraints.* Every relation should have a primary key and the value of the primary key in each tuple should be sole and cannot be null.
- *Referential integrity constraints.* Let a relation *r* have a foreign key *FK* and the foreign key value of a tuple *t* in *r* be *t* [*FK*]. Let *FK* quote the primary key *PK* of relation *r'* and *t'* be a tuple in *r'*. Referential integrity constraint demands that *t* [*FK*] comply with the following constraint: *t* [*FK*]= *t'* [*PK*]/null.
- *General integrity constraints.* In addition to the above-mentioned three kinds of integrity constraints that are most fundamental in relational database model, there are other integrity constraints related to data contents directly, called *general integrity constraints.* Because numbers of them are very large, only a few of them are considered in current relational DBMSs. Among these constraints, *functional dependencies* (*FD*) and *multivalued dependencies* (*MVD*) are more important in relational database design theory and widely investigated.

Operations. Relational database model provides some operations, called relational algebra operations. These operations can be subdivided into two classes: operations for relations only (select, project, join, and division) and set operations (union, difference, intersection, and Cartesian product). In addition, some new operations such as outerjoin, outerunion and aggregate operations are developed for database integration or statistics and decision support. By using these operations, one can query or update relations.

1.3.2 Nested Relational Database Model

The normalization, being one kind of constraints, is proposed in traditional relational databases. Among various normalized forms, first normal form (1NF) is the most fundamental one, which assumes that each attribute value in a relational instance must be atomic. As we know, the real-world applications are complex, and data types and their relationships are rich as well as complicated. The 1NF assumption limits the expressive power of traditional relational database model. Therefore, some attempts to relax 1NF limitation are made and one kind of data model, called non-first normal (or nested) relational database model have been introduced.

The first attempt to relax first-normal formal limitation is made by Makinouchi (1977), where, attribute values in the relation may be atomic or set-valued. Such relation is thereby called non-first normal form (NF^2) one. After Makinouchi's proposal, NF^2 database model is further extended (Ozsoyoglu *et al.*, 1987; Schek and Scholl, 1986). The NF^2 database model in common sense now means that attribute values in the relational instances are either atomic or set-valued and even relations themselves. So NF^2 databases are called nested relational databases also. In this paper, we do not differentiate these two notions. A formal definition of NF^2 relational schema is given as follows.

Definition. An attribute A_j is a structured attribute if its schema appears on the left-hand side of a rule; otherwise it is simple.

Definition. An NF^2 relational schema may contain any combination of simple or structured attributes on the right-hand side of the rules. Formally,

Schema:: Simple_attribute | Simple_attribute, Structured_attributes
Structured_attributes:: Simple_attribute | Simple_attribute, Structured_attributes

A schema is called flat if and only if all of its attributes are simple. It is clear that a classical schema, namely, a flat relational schema, is a special case of a nested relational schema. Two nested schemas are called union-compatible, meaning the ordered attributes have the same nesting structure, if and only if the corresponding simple attributes and structured attributes are union-compatible.

Let a relation *r* have schema $R = (A_1, A_2, ..., A_n)$ and let $D_1, D_2, ..., D_n$ be corresponding domains from which values for attributes $(A_1, A_2, ..., A_n)$ are selected. A tuple of an NF^2 relation is an element in *r* and denoted as $<a_1, a_2, ..., a_n>$ consisting of *n* components. Each component a_j $(1 \leq j \leq n)$ may be an atomic or null value or another tuple. If A_j is a structureed attribute, then the value a_j need not be a single value, but an element of the subset of the Cartesian product of associated domains $D_{j1}, D_{j2}, ..., D_{jm}$.

Let us look at the hierarchy structure of car products shown in Figure 1.5 (Erens *et al.*, 1994; Zhang and Li, 1999). The Car structure can be defined as a nested data model utilizing the following forms:

Car = (Car _Id, Interior, Chassis, Gearbox, Engine)
Interior = (Interior _Id, Dashboard, Seat)
Engine = (Engine_Id, Size, Turbo)

A NF^2 relational instance for the Car product data model is illustrated in Table 1.3.

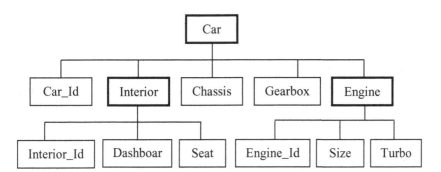

Fig. 1.5. Hierarchy structure of car product

Table 1.3. Nested relation for car product

Car_ Id	Interior			Chassis	Gear -box	Engine		
	Interior _Id	Dash_ board	Seat			Engine _Id	Size	Turbo
VA1	IN01	SE89	DB2	CH01	GE5	E18	1.8	No
VA2	IN02	SE35	DB3	CH02	GE5	E20	2.0	Yes

Based on the NF^2 database model, the ordinary relational algebra has been extended. In addition, two new restructuring operators, called the *Nest* and *Unnest* (Ozsoyoglu *et al.*, 1987; Roth *et al.*, 1987) (as well as *Pack* and *Unpack* (Ozsoyoglu *et al.*, 1987), have been introduced. The Nest operator can gain the nested relation including complex-valued attributes. The Unnest operator is used to flatten the nested relation. That is, it takes a relation nested on a set of attributes and desegregates it, creating a "flatter" structure. The formal definitions and the properties of these operators as well as the ordinary relational algebra for the NF^2 data model have been given (Colby, 1990; Venkatramen and Sen, 1993).

1.3.3 Object-Oriented Database Model

Although there has been great success in using the relational databases for transaction processing, the relational databases have some limitations in some non-transaction applications such as computer-aided design and manufacturing (CAD/CAM), knowledge-based systems, multimedia, and GIS. Such limitations include the following.

- The data type is very restricted.
- The data structure based on the record notion may not match the real-world entity.
- Data semantics is not rich, and the relationships between two entities cannot be represented in a natural way.

Therefore, some non-traditional data models were developed in succession to enlarge the application area of databases since the end of the 1970s. Since these non-traditional data models appeared after the relational data model, they are called post-relational database models. Object-oriented database model is one of the post-relational database models.

Object-oriented (OO) data model is developed by adopting some concepts of semantic data models and knowledge expressing models, some ideas of object-oriented program language and abstract data type in data structure/programming.

Objects and Identifiers

All real-world entities can be simulated as *objects*, which have no unified and standard definition. Viewing from the structure, an object consists of *attributes*, *methods* and *constraints*. The attributes of an object can be simple data and other objects. The procedure that some objects constitute a new object is called *aggregation*. A method in an object contains two parts: signature of the method that illustrates the name of the method, parameter type, and result type; implementation of the method.

In general, attributes, methods and constraints in an object are encapsulated as one unit. The state of an object is changed only by passing message between objects. *Encapsulation* is one of the major features in OO data models.

In OO data models, each object has a sole and constant identifier, which is called *object identifier* (OID). For two objects with same attributes, methods and constraints, they are different objects if they have a different OID. The OID of an object is generated by system and cannot be changed by the user.

The OID generated by system can be divided into two kinds, i.e., *logical object identifier* and *physical object identifier*. Logical object identifier is mapped into physical one when an object is used because only physical object identifier concerns the storage address of the object.

Classes and Instances

In OO data models, objects with the same attributes, methods and constraints can be incorporated into a *class*, where objects are called *instances*.

In a class, attributes, methods and constraints should be declared. Note that the attributes in a class can be classified into two kinds: instance variables and class variables. Instance variables are the attributes for which values are different in different objects of the class, while class variables are the attributes for which values are the same in different objects of the class.

In fact, classes can also be regarded as objects. Then, classes can be incorporated into another new class, called *meta class*. The instances of a meta class are classes. Therefore, objects are distinguished into *instance objects* and *class objects*.

Class Hierarchical Structure and Inheritance

A subset of a class, say *A*, can be defined as a class, say *B*. Class *B* is called a *subclass* and class *A* is called *superclass*. A subclass can further be divided into new subclasses. A *class hierarchical structure* is hereby formed, where it is permitted that a subclass has several direct or indirect superclasses. The relationship between superclass and subclass is called *IS-A relationship*, which represents a *specialization* from top to bottom and a *generalization* from bottom to top. Because one subclass can have several direct superclasses, a class hierarchical structure is not a tree but a class *lattice*.

Because a subclass is a subset of its superclass, the subclass inherits the attributes and methods in its all superclasses. Besides inheritance, a subclass can define new attributes and methods or can modify the attributes and methods in the superclasses. If a subclass has several direct superclasses, the subclass inherits the attributes and methods from these direct superclasses. This is called *multiple inheritance*.

When inheriting, the naming conflict may occur, which should be resolved.

- Conflict among superclasses. If several direct superclasses of a subclass have the same name of attributes or methods, the conflict among superclasses appear. The solution is to declare the superclass order inherited, or to be illustrated by user.
- Conflict between a superclass and a subclass. When there are conflicts between a subclass and a superclass, the definition of attributes and methods in subclass would override the same definition in the superclass.

Note that a naming method may have a different meaning in different classes. The feature that a name has a multiple meaning is called *polymorphism*. The method with polymorphism is called *overloading*. Because the method in an object is polymorphism, the procedure corresponding to the method name cannot be determined while compiling program and do while running program. The later combination of the method name and implementing procedure of a method is called *late binding*.

1.4 Constructions of Database Models

Depending on data abstract levels and actual applications, different database models have their advantages and disadvantages. This is the reason why there exist a lot of database models, conceptual ones and logical ones. It is not appropriate to state that one database model is always better than the others. Conceptual data models are generally used for engineering information modeling at a high level of abstraction. However, engineering information systems are constructed based on logical database models. So at the level of data manipulation, i.e., a low level of abstraction, logical database model is used for engineering information modeling. Here, logical database models are often created through mapping conceptual data models into logical database models. This conversion is called *database conceptual design*. The relationships among conceptual data models, logical database models, and engineering information systems are shown in Figure 1.6.

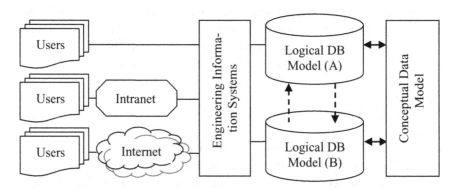

Fig. 1.6. Relationships among conceptual data model, logical database model, and engineering information systems

In this figure, *Logical DB Model (A)* and *Logical DB Model (B)* are different database systems. That means that they may have different logical database models, say relational database and object-oriented database, or they may be different database products, say *Oracle* and *DB2*, although they have the same logical database model. It can be seen from the figure that a developed conceptual data model can be mapped into different logical database models. Besides, it can also be seen that a logical database model can be mapped into a conceptual data model. This conversion is called *database reverse engineering*. It is clear that it is possible that different logical database models can be converted each other through database reverse engineering.

1.4.1 Development of Conceptual Data Models

It has been shown that database modeling of engineering information generally starts from conceptual data models and then the developed conceptual data models are mapped into logical database models. First of all, let us focus on the choice, design, conversion, and extension of conceptual data models in database modeling of engineering information.

Generally speaking, ER and IDEF1X data models are good candidates for business process in engineering applications. But for design and manufacturing, object-oriented conceptual data models such EER, UML, and EXPRESS are powerful. Being the description methods of STEP and a conceptual schema language, EXPRESS is extensively accepted in industrial applications. However, EXPRESS is not a graphical schema language, unlike EER and UML. In order to construct EXPRESS data model at a higher level of abstract, EXPRESS-G, being the graphical representation of EXPRESS, is introduced. Note that EXPRESS-G can only express a subset of the full language of EXPRESS. EXPESS-G provides supports the notions of entity, type, relationship, cardinality, and schema. The functions, procedures, and rules in EXPRESS language are not supported by EXPRESS-G. So EER and UML should be used to design EXPRESS data model conceptually and then such EER and UML data models can be translated into EXPRESS data model.

That multiple graphical data models can be employed facilitates the designers with different background to design their conceptual models easily by using one of the graphical data models they are familiar with. However, a complex conceptual data model is generally completed cooperatively by a design group, in which each member may use a different graphical data model. All these graphical data models designed by different members should be converted into a union data model finally. So far, the data model conversions among EXPRESS-G, IDEF1X, ER/EER, and UML only receive few attentions although such conversions are crucial in engineering information modeling. In (Cherfi *et al.*, 2002), the conceptual modeling quality between EER and UML was investigated. In (Arnold and Podehl, 1999), a mapping from EXPRESS-G to UML was introduced in order to define a linking bridge and bring the best of worlds of product data technology and software engineering together. Also the formal transformation of EER and EXPRESS-G was developed in (Ma *et al.*, 2003). In addition, the comparison of UML and IDEF was given in (Noran, 2000).

Figure 1.7 shows the design and conversion relationships among conceptual data models.

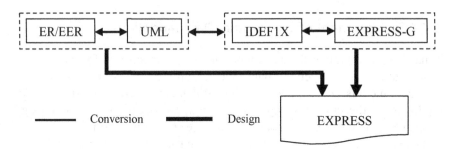

Fig. 1.7. Relationships among conceptual data models

1.4.2 Development of Logical Database Models

It should be noticed that there might be semantic incompatibility between conceptual data models and logical database models. So when a conceptual data model is mapped into a logical database model, we should adopt such a logical database model which expressive power is close to the conceptual data model so that the original information and semantics in the conceptual data model can be preserved and supported furthest. Table 1.4 shows how relational and object-oriented databases fair against various conceptual data models. Here, "*CDM*" and "*LDBM*" denote conceptual data model and logical database model, respectively.

Table 1.4. Match of logical database models to conceptual data models

CDM \ LDBM	Relational Databases	Object-Oriented Databases
ER	*good*	*bad*
IDEF1X	*good*	*bad*
EER	*fair*	*good*
UML	*fair*	*good*
EXPRESS	*fair*	*good*

It is clear from the table that relational databases support ER and IDEF1X well. So, when an ER or IDEF1X data model is converted, relational databases should be used. It is also seen that EER, UML, or EXPRESS data model should be mapped into object-oriented databases. EXPRESS is extensively accepted in industrial application area. EER and UML, being graphical conceptual data models, can be used to design EXPRESS data model conceptually and then EER and UML data models can be translated into EXPRESS data model (Oh *et al.*, 2001).

Concerning logical database implementation of EXPRESS data model, the following tasks must be performed:

- defining the database structures from EXPRESS data model and
- providing SDAI (STEP Standard Data Access Interface) access to the database.

Users define their databases using EXPRESS model, manipulate the databases using SDAI, and exchange data with other applications through the database systems. The implementation of SDAI/STEP will be given in Chapter 10.

1.5 Summary

Engineering information modeling is one of crucial tasks to implement engineering information systems. Databases are designed to support data storage, processing, and retrieval activities related to data management, and database systems are the key to implementing engineering information modeling. Database models can be classified into conceptual data models and logical database models. The purpose of engineering information modeling in databases is to construct the logical database models, which are the foundation of the engineering information systems. Generally the constructions of logical database models start from the constructions of conceptual data models and then the developed conceptual data models are converted into the logical database models.

It should be noticed that, however, the current mainstream databases are mainly designed for business applications. Engineering information modeling is complex because it should cover product life cycle times and engineering applications are typically data and knowledge intensive applications. There are some unique requirements from engineering information modeling, which impose a challenge to databases technologies and promote their evolvement. It is especially true for contemporary engineering applications, where some new techniques have been increasingly applied and their operational patterns are hereby evolved (e.g., e-manufacturing, Web-based PDM, and etc). In (Ma, 2005), some requirements for engineering information modeling, including *complex objects and relationships, data exchange and share, Web-based applications, imprecision and uncertainty,* and *knowledge management,* have been identified. Among these requirements, the imprecision and uncertainty widely exist in engineering applications and have been investigated in the context of various engineering activities. For example, imprecision is most notable in the early phase of the design process and product design is a process of reducing the imprecision in the description of the conceptual design (Antonsoon and Otto, 1995). Currently, there are no existing modeling tools and database systems supporting imprecise and uncertain engineering information modeling. This imposes great challenges to current database technologies in modeling imprecise and uncertain engineering information. Although there have been some attempts to resolve the problems involving information imprecision and uncertainty, until now, little has been done in the field for modeling imprecise and uncertain engineering information in databases.

References

Ambler, S. W. (2000a), The design of a robust persistence layer for relational databases, http://www.ambysoft.com/persistenceLayer.pdf.

Ambler, S. W. (2000b), Mapping objects to relational databases, http://www.AmbySoft.com/mappingObjects.pdf.

Antonsoon, E. K. and Otto, K. N. (1995), Imprecision in engineering design, ASME Journal of Mechanical Design, 117 (B): 25-32.

Appleton (1986), Information Modeling Manual–IDEF1Extended (IDEF1X), D. Appleton Company, Manhattan Beach, CA.

Arnold, F. and Podehl, G. (1999), Best of both worlds–a mapping from EXPRESS-G to UML, Lecture Notes in Computer Science 1618, 49-63.

Blaha, M. and Premerlani, W. (1999), Using UML to design database applications, http://www.therationaledge.com/rosearchitect/mag/archives/9904/f8.html.

Booch, G., Rumbaugh, J. and Jacobson, I. (1998), The unified modeling language user guide, Addison-Welsley Longman, Inc.

Chen, P. P. (1976), The entity-relationship model: toward a unified view of data, ACM Transactions on Database Systems, 1 (1): 9-36.

Cherfi, S. S. S., Akoka, J. and Comyn-Wattiau, I. (2002), Conceptual modeling quality – from EER to UML schemas evaluation, Lecture Notes in Computer Science 2503, 414-428.

Codd, E.F. (1970), A relational model of data for large shared data banks, Communications of the ACM, 13 (6): 377-387.

Conrad, R., Scheffner, D. and Freytag, J. C. (2000), XML conceptual modeling using UML, Lecture Notes in Computer Science 1920, 558-571.

dos Santos, C., Neuhold, E. and Furtado, A. (1979), A data type approach to the entity-relationship model, Proceedings of the 1st International Conference on the Entity-Relationship Approach to Systems Analysis and Design, 103-119.

Eastman, C. M. and Fereshetian, N. (1994), Information models for use in product design: a comparison, Computer-Aide Design, 26 (7): 551-572.

Elmasri, R., Weeldreyer, J. and Hevner, A. (1985), The category concept: an extension to the entity-relationship model, International Journal on Data and Knowledge Engineering, 1 (1): 75-116.

Gegolla, M. and Hohenstein, U. (1991), Towards a semantic view of an extended entity-relationship model, ACM Transactions on Database Systems, 16 (3): 369-416.

IDEF (2000), IDEF Family of Methods, http://www.idef.com/default.html

ISO IS 10303-1 TC184/SC4 (1994), Product Data Representation and Exchange-Part 1: Overview and Fundamental Principles, International Standard, 1994.

ISO IS 10303-1 TC184/SC4 (1994), Product Data Representation and Exchange-Part 11: The EXPRESS Language Reference Manual, International Standard.

Kusiak, A., Letsche, T. and Zakarian, A. (1997), Data modeling with IDEF1X, International Journal of Computer Integrated Manufacturing, 10: 470-486.

Ma, Z. M. (2005), Database modeling of engineering information: needs and constructions, Industrial Management and Data Systems, 105 (7): 900-918.

Ma, Z. M., Lu, S. Y. and Fotouhi, F. (2003), Conceptual data models for engineering information modeling and formal transformation of EER and EXPRESS-G, Lecture Notes in Computer Science 2813, 573-575.

Makinouchi, A. (1977), A consideration on normal form of not-necessarily normalized relations in the relational data model, Proceedings of Third International Conference on Very Large Databases, 447-453.

Männistö, T., Peltonen, H., Soininen, T. and Sulonen, R. (2001), Multiple abstraction levels in modeling product structures, Date and Knowledge Engineering, 36 (1): 55-78.

McKay, A., Bloor, M. S., de Pennington, A. (1996), A framework for product data, IEEE Transactions on Knowledge and Data Engineering, 8 (5): 825-837.

Mili, F., Shen, W., Martinez, I., Noel, Ph., Ram, M. and Zouras, E. (2001), Knowledge modeling for design decisions, Artificial Intelligence in Engineering, 15: 153-164.

Naiburg, E. (2000), Database modeling and design using rational rose 2000, http://www.therationaledge.com/rosearchitect/mag/current/spring00/f5.html.

Noran, O. (2000), Business Modeling: UML vs. IDEF, Report/Slides, School of CIT, Griffith University, www.cit.gu.edu.au/~noran.

OMG (2003), Unified modeling language (UML), http://www.omg.org/technology/documents/formal/uml.htm.

Ozsoyoglu, G., Ozsoyoglu, Z. M. and Matos, V. (1987), Extending relational algebra and relational calculus with set-valued attributes and aggregate functions, ACM Transactions on Database Systems, 12 (4): 566-592.

Roth, M. A., Korth, H. F. and Batory, D. S. (1987), SQL/NF: A query language for non-1NF relational databases, Information Systems, 12: 99-114.

Schek, H. J. and Scholl, M. H. (1986), The relational model with relational-valued attributes, Information Systems, 11 (2): 137-147.

Schenck, D. A. and Wilson, P. R. (1994), Information Modeling: the EXPRESS Way, Oxford University Press.

Scheuermann, P., Schiffner, G. and Weber, H. (1979), Abstraction capabilities and invariant properties modeling within to the entity-relationship approach, Proceedings of the 1st International Conference on the Entity-Relationship Approach to Systems Analysis and Design, 121-140.

Shaw, N. K., Bloor, M. S. and de Pennington, A. (1989), Product data models, Research in Engineering Design, 1: 43-50.

Softech Inc. (1981), Information modeling manual (IDEF1), USAF Integrated Computer Aided Manufacturing Program, AFWAL-TR-81-4023, Wright-Patterson AFB, OH.

Wizdom Systems, Inc. (1985), U.S. Air Force ICAM Manual: IDEF1X, Naperville, IL, November.

Zhang, W. J. and Li, Q. (1999), Information modeling for made-to-order virtual enterprise manufacturing systems, Computer-Aided Design, 31 (10): 611-619.

2 Information Imprecision and Uncertainty in Engineering

2.1 Introduction

As it may be known, in real-world applications, information is often vague or ambiguous. Therefore, different kinds of imperfect information have extensively been introduced and studied to model the real world as accurately as possible.

Information-based engineering applications are typically data and knowledge intensive applications. There are some unique requirements from engineering information modeling. In addition to complex structures and rich semantic relationships, one also needs to model imprecise and uncertain information in many engineering activities. Information imprecision and uncertainty exist in almost all engineering applications and has been investigated in the context of various engineering actions.

2.2 Imprecision and Uncertainty in Engineering

Imprecision is most notable in the early phase of the design process and has been defined as the choice between alternatives (Antonsoon and Otto, 1995). Four sources of imprecision found in engineering design have been classified as *relationship imprecision*, *data imprecision*, *linguistic imprecision*, and *inconsistency imprecision* in (Giachetti *et al.*, 1997). In addition to engineering design, imprecise and uncertain information can be found in many engineering activities. The imprecision and uncertainty in activity control for product development was investigated in (Grabot and Geneste, 1998). A design framework for PAC (production activity control) decision support systems was hereby developed. To manage the uncertainty occurring in industrial firms, various types of buffers were provided in (Caputo, 1996) according to different types of uncertainty faced and to the characteristics of the production system. Buffers are used as alternative and

Z. Ma: *Fuzzy Database Modeling of Imprecise and Uncertain Engineering Information*,
StudFuzz **195**, 33–45 (2006)
www.springerlink.com

complementary factors to attain technological flexibility when a firm is unable to achieve the desired level of flexibility and faces uncertainty. Nine types of flexibility (*machine, routing, material handling system, product, operation, process, volume, expansion,* and *labor*) in manufacturing were summarized in (Tsourveloudis and Phillis, 1998).

2.2.1 Product Preliminary Design

Product design is generally divided into three stages (Hsu and Woon, 1998), i.e., product design specification, conceptual design, and detailed design. The information of the product is collected and defined during the first stage. Physical solutions are generated to meet the design specification at the second stage. The final decisions on the dimensions, layout and shape of individual components and material to be adopted are made at the final stage. The relationship among these stages is illustrated in Figure 2.1.

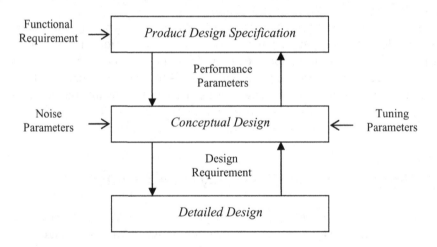

Fig. 2.1. Phases of product design

Conceptual design performed in the early stages of product development has a significant influence on such factors as cost, performance, reliability, safety, and environmental impact of the final product. Conceptual design is therefore a complex process. On the one hand, a product should be described by its function, behaviour and structure (Hsu and Woon, 1998; Zhang *et al.*, 1997). On the other hand, it is difficult to obtain the complete and exact information/knowledge such as design requirements and constraints in the initial phase of product design. Product design is, in fact, a process of reducing the incompleteness in the description of conceptual design.

In the literature, one can find a Method of Imprecision (MoI) to perform design calculation on imprecise representation of parameters based on fuzzy set theory (Otto and Antonsoon, 1994a; Wood and Antonsoon, 1992). With MoI, customers can provide fuzzy requirements of performance with certain preferences. Engineering designers also assign fuzzy design values with certain preferences following the requirement of the customers. The final design should be the trade-off between two aspects. In addition to MoI, one can also find approaches for representing and manipulating imprecision (Antonsoon and Otto, 1995). These approaches include utility theory, optimization, matrix method, probability method, and necessity method and their advantage and disadvantage have also been addressed.

In addition to design and performance requirements, factors at other stages of the product life cycle, such as manufacturing and assembly should also be considered in product conceptual design. Four kinds of parameters are identified at the stages of product conceptual design (Otto and Antonsoon, 1994b):

- design parameters,
- performance parameters,
- noise parameters, and
- tuning parameters.

Performance parameters are the requirements of the final product given by the customer, such as stress requirement. Design parameters are those assigned by designers, such as geometric sizes. When a designer chooses design parameters, it should be guaranteed that the performance parameters could satisfy the requirements. Some noises, such as manufacturing errors, should also be considered to improve the robustness of the product. It is worthy to mention that, although the noise parameters are considered before assigning design parameters, the final design may sometimes not satisfy the performance requirements. Tuning parameters are thus introduced into engineering design, which can be used to overcome the confounding influences of the noise parameters. Tuning parameters are often assigned in the manufacturing phase based on the confounding influences of the noise. Designers do not set the tuning parameters, but they should take the tuning parameters into account when assigning design parameters.

It can be seen that these four kinds of parameters have different characteristics and their values are determined using different mechanisms. Being similar to design and performance parameters, noise and tuning parameters are also in general imprecise or uncertain in the product conceptual design phase. In the following, the imprecise and uncertain semantics included in these parameters are illustrated.

Imprecision in Product Preliminary Design

An imprecise variable corresponds to a range or a set of values, in which only one will be true. The imprecision in product conceptual design includes two levels of meanings. For the universe of continua discourse, the design parameters or performance parameters are usually represented by a range as well as a function which is defined on that range. For the universe of discrete discourse, however, the design parameters or performance parameters are usually represented by a set of values as well as the preferences corresponding to each value. The function value or the preference corresponding to a value in the range or the set of values indicates the possibility that this value is true. Such imprecision in design and performance parameters can be represented by fuzzy values.

Suppose that for a product, the length parameter L is described by "about 150 cm" and the volume parameter V is described by "no more than 10 l", in which L and V are the design parameter and the performance parameter, respectively. For instance, the imprecise data "no more than 10 l" may be represented by a set of values such as {1.0/1, 1.0/2, 1.0/3, 1.0/4, 1.0/5, 0.9/6, 0.8/7, 0.6/8, 0.3/9}, in which each value is coupled with a preference.

Tuning parameters are set by manufacturers in the manufacturing phase or even by customers based on the effect of noises. Although a designer does not know the precise values of tuning parameters, one knows that the values must be one of a set of values or within an interval of values. Therefore, tuning parameters should also be represented by a range or a set of values in the phase of product conceptual design. Different from the design parameters and performance parameters, for a tuning variable, each value in the range or the set of values has an equal possibility of being true. Their possibilities are all 1. Designers do not need to provide the preferences for tuning parameters. An example of imprecise tuning parameter, for example, is represented by an interval [5, 10].

Uncertainty in Product Preliminary Design

Product design should be manufacturing oriented. Noise parameters should, therefore, also be considered in the product conceptual design stage. The characteristics of noise parameters are uncontrolled and stochastic. For example, the probability that the manufacturing tolerance of length x is 0.9 mm is 0.7. Such characteristics that noise parameters have values indicate that they are stochastic and should be modeled using probabilistic values. A noise variable can be represented by a probabilistic distribution for the universe of discrete discourse, or a distribution function for the universe of continua discourse.

2.2.2 R2R Controller Implementation

At the run-to-run (R2R) control level, control is carried out by utilizing data from the results of the last run, or a series of previous runs for deriving better inputs for the next run. The R2R controller uses a database to store the results of the process. In certain cases it is difficult to get accurate measurements of the results of a process (i.e., metrology data). The ability to store fuzzy data allows the R2R controller to handle such situations. Therefore, the R2R controller for semiconductor control is developed in (Chaudhry *et al.*, 1999) for handling fuzzy data (see Figure 2.2).

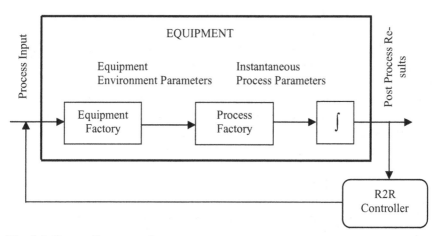

Fig. 2.2. Run-to-Run control

On the other hand, one of the tasks of the R2R controller is the selection of control algorithms for optimization/control of the process. After each run of the process, the run data is given as input to the R2R controller. After performing some checks to ascertain whether the process can be optimized or not, the controller has to decide which algorithm or set of algorithms should be invoked to optimize or control the process. Generally, there is no single algorithm that can be used throughout the entire range of R2R control and optimization. The range of applicability of these algorithms can be roughly expressed in terms of the process being near or far from its optimum. It is important to exploit the knowledge, even if vague, to assist the selection of the control algorithms. In addition, many rules that a process engineer might have for process control are most likely expressed in terms of a natural language with many vague and imprecise terms.

2.2.3 Production Activity Control (PAC)

The operational levels of production management are often called product activity control (PAC) or manufacturing process control. PAC requires increasing reaction capabilities in order to adapt the workshop management to changes in its environment (Grabot and Geneste, 1998). PAC can be defined as a group of activities directly in charge of managing the transformation of planned orders into a set of outputs. It governs the very short-term and detailed planning, execution, and monitoring activities required to control the flow of an order from the time when the order is released by the planning system for execution until that order is filled and its disposition completed (Melnyk and Carter, 1986).

The PAC system applies to middle-term decision making of the upper levels with the adaptations required by short-term or real-time disturbances and it is essential to ensure reactivity along with optimization of resource use. It is important to use the middle-term decision making with as few modifications as possible since these decisions aim at optimizing the use of resources (e.g. by load smoothing). In PAC, four main functions can be identified (Huguet and Grabot, 1995).

- Function plan considers the inputs of the production plan and the objectives transmitted by the parent decision center. It implements techniques to adapt the production plan to the stated objectives and makes it operational on a fixed time horizon and period.
- Function launch decomposes, synchronizes, and executes decisions taken within the PAC module.
- Function follow-up collects all the results provided by sensors on decision modules as well as the requests external to the shopfloor, processes them according to their types and aggregation levels, and transmits them to other functions and PAC modules.
- Function react allows a real-time adaptation of the decisions taken at the different management levels. It acts punctually by adapting a plan that is not directly feasible. It can also re-launch the function plan after a disturbance.

Figure 2.3 shows a PAC module that can be defined using these functions. It can be implemented at various decision levels as a macro-decision center.

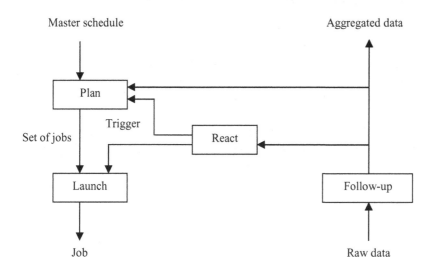

Fig. 2.3. Production activity control module

The main decisions made at the PAC level are as follows:

- At the planning level (mainly load planning or scheduling): choice of the resource calendars, choice of resources in order to perform an operation (when there are replacement resources), and starting date of each operation.
- At the launching level: resource permutations, manufacturing order interruptions, and manufacturing order permutations on a resource.
- At the follow-up level: there is no real decision making, but the degree of satisfaction is evaluated as given by the current workshop behavior.
- At the reaction level: decision-making is triggered, when events do not allow the objective satisfaction occurs anymore, or when the discrepancy between the real objective satisfaction and the required satisfaction is too high. Solutions may be applied at two levels, planning and launching.

The design of decision support systems at the PAC level first aims at taking into account high-level decision criteria in operational levels. This de facto modification of the decision structure requires efficient tools for sorting possible solutions in accordance with the satisfaction of the mid-term workshop objectives. A substantial constraint on such a tool is that it should be able to manage information often pervaded by imprecision and uncertainty at the PAC level.

Imprecision and Uncertainty in PAC Systems

A general model of a PAC decision center results in imprecision and uncertainty at four levels (Grabot and Geneste, 1998).

- The decision trigger (i.e., whether or not a decision should be made when an unexpected event occurs).
- The reasoning that allows the decision to be made.
- The technical data used by this reasoning.
- The workshop objectives that the final decisions should satisfy.

Figure 2.4 shows the relationship between imprecision and decision making.

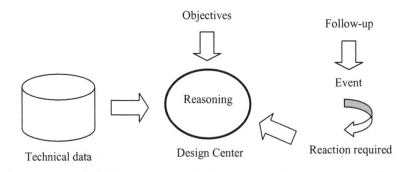

Fig. 2.4. Imprecision and decision making

At various levels, the decisions are usually carried out for a given period of time when no serious disturbance occurs. In addition, a decision may also be triggered when the difference between required and real performance reaches a given threshold. It could also be triggered if the accumulation of little disturbances exceeds a given threshold, which would otherwise not require any decision/action if appeared individually. The threshold depends on the manufacturing context. It is difficult to define a priori, and instead it is useful to express the threshold with imprecision. The reasoning can then be adapted to several close contexts.

The elementary reasoning processes used for making other decisions may propagate imprecision and/or uncertainty to the conclusions. Various cases can be considered as follows:

- Proportional or reverse proportional reasoning: A classical dispatching rule, e.g. states that the later a manufacturing order, the higher its priority. This kind of reasoning may be understood as an imprecision on the conditions inducing an imprecision on the reasoning conclusion.

- The imprecision on the conditions may also induce an uncertainty on the conclusion, when the validity of reasoning is no longer certain since the real context becomes far from the standard context. For instance, an imprecise evaluation of the workshop performance may induce an uncertainty on the need to trigger the react function.

These two kinds of reasoning may be combined, i.e., an imprecision on the conditions may induce both an imprecision (because of the approximate reasoning) and an uncertainty (due to the validity limit of the reasoning). For instance, a resource may become a bottleneck during an overload. In this case it is interesting to decrease the load of the resource by using a replacement resource or an increase in the batch size. The more substantial the overload, the more substantial the increase in batch size. When there is an imprecise load rate, perhaps due to an imprecision of the orders, these solutions become not only imprecise but also uncertain, since it is no longer certain that the resource is a bottleneck.

An uncertainty on a condition always induces at least an equivalent uncertainty on the conclusion. Most of the time, the complex reasoning processes are simplified by procedures using thresholds or arbitrary intervals in order to be easily transmitted to different operators. Manufacturing data are pervaded by imprecision or uncertainty, e.g., the forecast orders that allow us to build the master production plan. Another example is the average duration of an operation (including all kinds of waiting time) used in the MRP (Materials Requirements Planning) method for carrying out backward scheduling. On the basis of the delivery dates, this average duration allows us to determine roughly the period over which the manufacturing order uses each resource, and to detect possible overloads. Load smoothing then enables us to determine the starting dates of the operations in precise scheduling. But the real duration of the operations is only known after building the schedule. The introduced imprecision results from an arbitrary choice made early on, and it could be subject to question at the end of the reasoning phase. In fact, the great volume of processed information does not allow us to make loops until the reasoning converges.

Another imprecision is ignorance of an operation's duration, perhaps because it is a manual operation or because the length of time is difficult to forecast, e.g., for a maintenance operation. This imprecision in the duration is naturally the cause of an imprecision in the resource capacity, and resource capacity is used when a load/capacity adequation is required.

Data linked to supplies are also pervaded by imprecision, such as the supply delay or the delivered quantity. These kinds of imprecision are usually dealt with:

- probabilities, e.g., supplies: computing the safety stock for protection from disturbances can be based on a protection factor equal to the appropriate number of standard deviations in the consumption, and
- an increase in quantities or delays in order to be protected from disturbances (e.g., increase of batch sizes or manufacturing delays).

The first solution requires us to gather a sufficient set of data in order to be efficient, whereas the second one is only a last resort that increases the amount of work in progress and follow-up difficulties.

Being a convenient tool for knowledge modeling, fuzzy logic should also be used in addition to expert systems in order to manage imprecise and uncertain information.

2.3 Representation of Imprecise and Uncertain Engineering Information

Concerning the representation of imprecision and uncertainty, attempts have been made to address this issue in design by way of intervals (Kim *et al.*, 1995). Other approaches to representing imprecision in design include using utility theory, implicit representations using optimization methods, matrix methods such as Quality Function Deployment, probability methods, and necessity methods. An extensive review of these approaches was provided in (Antonsoon and Otto, 1995). These methods have all had limited success in solving design problems with imprecision. It is believed that fuzzy reorientation of imprecision will play an increasingly important role in design systems (Zimmermann, 1999).

Fuzzy set theory (Zadeh, 1965) is a generalization of classical set theory. In normal set theory, an object may or may not be a member of a set. There are only these two states. Fuzzy sets contain elements to a certain degree. Thus, it is possible to represent an object that has partial membership in a set. The membership value of element u in a fuzzy set is represented by $\mu(u)$ and is normalized such that $\mu(u)$ is in [0, 1].

Fuzzy sets can represent linguistic terms and imprecise quantities and make systems more flexible and robust. Therefore, fuzzy set theory has been used in some engineering applications (e.g., engineering/product design and manufacturing, production management, manufacturing flexibility, e-manufacturing, and etc.), where either crisp information is not available or information-flexible processing is necessary.

Concerning engineering/product design and manufacturing, the needs for fuzzy logic in the development of CAD systems were identified and how fuzzy logic can be used to model aesthetic factors was discussed in

(Pham, 1998). The development of an expert system with production rules and the integration of fuzzy techniques (fuzzy rules and fuzzy data calculus) was described for the preliminary design in (Francois and Bigeon, 1995). Integrating knowledge-based methods with multi-criteria decision making and fuzzy logic, an approach to engineering design and configuration problems was developed in order to enrich existing design and configuration support systems with more intelligent abilities in (Muller and Sebastian, 1997). A methodology for making the transition from imprecise goals and requirements to the precise specifications needed to manufacture the product was introduced using fuzzy set theory in (Giachetti *et al.*, 1997). In (Jones and Hua, 1998), an approach to engineering design in which fuzzy sets were used to represent the range of variants on existing mechanisms was described so that novel requirements of engineering design can be met. A method for design candidate evaluation and identification using neural network-based fuzzy reasoning was presented in (Sun *et al.*, 2000).

In production management, the potential applications of fuzzy set theory to new product development, facility location and layout, production scheduling and control, inventory management, and quality and cost-benefit analysis were identified in (Karwowski and Evans, 1986). A comprehensive literature survey on fuzzy set applications in product management research was given in (Guiffrida and Nagi, 1998). A classification scheme for fuzzy applications in product management research was defined in their paper, including job shop scheduling, quality management, project scheduling, facilities location and layout, aggregate planning, production and inventory planning, and forecasting.

In the manufacturing domain, flexibility is an inherently vague notion. Therefore, fuzzy logic was introduced and a fuzzy knowledge–based approach was used to measure manufacturing flexibility (Tsourveloudis and Phillis, 1998).

More recently, the research on supply chain management and electronic commerce has also shown that fuzzy set theory can be used in determining customer demand, supply deliveries along the supply chain, external or market supply, targeted marketing, and product category description (Petrovic *et al.*, 1998 & 1999; Yager, 2000; Yager and Pasi, 2001). It is believed that fuzzy set theory has considerable potential for intelligent manufacturing systems and will be employed in more and more engineering applications.

2.4 Summary

Several engineering activities in which imprecise and uncertain information can be found have been demonstrated in the chapter. In fact, information imprecision and uncertainty exist almost in all aspects of engineering applications, such as industrial controllers, production management, expert systems, and optimization and operation research. Therefore, there have been some attempts to represent imprecise and uncertain engineering information. Among various methods for the representation of imprecision and uncertainty in engineering, it has been believed that fuzzy set theory put forward by L. A. Zadeh in 1965 (Zadeh, 1965) can play an important role. And fuzzy set theory has been used in many engineering applications.

The next chapter will brief present fuzzy set theory.

References

Antonsoon, E. K. and Otto, K. N. (1995), Imprecision in engineering design, ASME Journal of Mechanical Design, 117 (B): 25-32.

Caputo, M. (1996), Uncertainty, flexibility and buffers in the management of the firm operating system, Production Planning & Control, 7 (5): 518-528.

Chaudhry, N. A., Moyne, J. and Rundensteiner, E. A. (1999), An extended database design methodology for uncertain data management, Information Sciences, 121 (1-2): 83-112.

Francois, F. and Bigeon, J. (1995), Integration of fuzzy techniques in a CAD/CAM system, IEEE Transactions on Magnetics, 31 (3): 1996-1999.

Giachetti, R. E., Young R. E., Roggatz, A., Eversheim W. and Perrone G. (1997), A methodology for the reduction of imprecision in the engineering design process, European Journal of Operations Research, 100 (2): 277-292.

Grabot, B. and Geneste, L. (1998), Management of imprecision and uncertainty for production activity control, Journal of Intelligent Manufacturing, 9: 431-446.

Grabot, B. and Geneste, L. (1998), Management of imprecision and uncertainty for production activity control, Journal of Intelligent Manufacturing, 9: 431-446.

Guiffrida, A. and Nagi, R. (1998), Fuzzy set theory applications in production management research: a literature survey, Journal of Intelligent Manufacturing, 9: 39-56.

Hsu, W. and Woon, I. M. Y. (1998), Current research in the conceptual design of mechanical products, Computer-aided Design, 30 (5): 377-389.

Huguet, P. and Grabot, B. (1995), A conceptual framework for shopfloor production activity control, International Journal of Computer Manufacturing, 8 (5): 357-359.

Jones, J. D. and Hua, Y. (1998), A fuzzy knowledge base to support routine engineering design, Fuzzy Sets and Systems, 98 (3): 267-278.

Karwowski, W. and Evans, G. W. (1986), Fuzzy concepts in production management research: a review, International Journal of Production Research, 24 (1): 129-147.

Kim, K., Cormier, D., O'Grady, P. and Young, R. E. (1995), A system for design and concurrent engineering under imprecision, Journal of Intelligent Manufacturing, 6 (1): 11-27.

Melnyk, S. A. and Carter, P. L. (1986), Identifying the principles of effective production activity control, Proceedings of the 29th International Conference of American Production and Inventory Control Society, 227-232.

Muller, J. and Smith, G. (1993), A pre-competitive project in intelligent manufacturing technology, Proceedings of AAAI '93 Workshop on Intelligent Manufacturing Technology.

Otto, K. N. and Antonsoon, E. K. (1994a), Modeling imprecision in product design, Proceedings of Fuzzy-IEEE 1994, 346-351.

Otto, K. N. and Antonsoon, E. K. (1994b), Design parameter selection in the presence of noise, Research in Engineering Design, 6 (4): 234-246.

Petrovic, D., Roy, R. and Petrovic, R. (1998), Modeling and Simulation of a Supply Chain in an Uncertain Environment, European Journal of Operational Research, 109: 299-309.

Petrovic, D., Roy, R. and Petrovic, R. (1999), Supply chain modeling using fuzzy sets, International Journal of Production Economics, 59: 443-453.

Pham, B. (1998), Fuzzy logic applications in computer aided design, Fuzzy Systems Design, Studied in Fuzziness and Soft Computing 17, 73-85.

Sun, J., Kalenchuk, D. K., Xue, D. and Gu, P. (2000), Design candidate identification using neural network-based fuzzy reasoning, Robotics and Computer Integrated Manufacturing, 16 (5): 383-396.

Tsourveloudis, N. G. and Phillis, Y. A. (1998), Manufacturing flexibility measurement: a fuzzy logic framework, IEEE Transactions on Robotics and Automation, 14 (4): 513-524.

Wood, K. L. and Antonsoon, E. K. (1992), Modeling imprecision and uncertainty in preliminary engineering design", Mechanism and Machine Theory, 25 (3): 305-324.

Yager, R. R. (2000), Targeted e-commerce marketing using fuzzy intelligent agents, IEEE Intelligent Systems, 15 (6): 42-45.

Yager, R. R. and Pasi, G. (2001), Product category description for web-shopping in e-commerce, International Journal of Intelligent Systems, 16: 1009-1021.

Zadeh, L. A. (1965), Fuzzy sets, Information and Control, 8 (3): 338-353.

Zhang, W. J., Zhang, D. and van der Werff, K. (1997), Toward an integrated data representation of function, behavior and structure for computer aided conceptual mechanical system design, Integrated Product and Process Development: Methods, Tools and Technologies, John Wilery & Sons, 85-124.

Zimmermann, H. J. (1999), Practical Applications of Fuzzy Technologies, Kluwer Academic Publishers.

3 Fuzzy Sets and Possibility Distributions

3.1 Introduction

Fuzzy set theory (Zadeh, 1965), which is interchangeably referred as fuzzy logic, is a generalization of the set theory and provides a means for the representation of imprecision and vagueness. In real-world applications, information is often imperfect. Therefore, fuzzy set theory has been applied in a number of real applications crossing over a broad realm of domains and disciplines (Cox, 1995; Munakata and Jani, 1994) since its formal start in 1965. Currently fuzzy logic (FL) along with neural networks (NN), genetic algorithms (GA), probabilistic reasoning (PR), chaos theory (ChT) and etc. has been main components of Soft Computing (Aliev *et al.*, 2001).

The purpose of this chapter is to review the basic definitions and results on fuzzy sets and related topics.

3.2 Imperfect Information

In real-world applications, information is often vague or ambiguous. There have been some attempts at classifying various possible kinds of imperfect information. Inconsistency, imprecision, vagueness, uncertainty, and ambiguity are five basic kinds of imperfect information in database and information systems (Bosc and Prade, 1993; Motor and Smets, 1997).

Inconsistency is a kind of semantic conflict when one aspect of the real world is irreconcilably represented more than once in a data set (database) or several different data sets (databases). For example, one has "*married*" value and "*single*" value for Tom's marital status. Information inconsistency usually comes from information integration.

Intuitively, imprecision and vagueness are relevant to the content of an attribute value, and it means that a choice must be made from a given range (interval or set) of values but it is not known exactly which one to choose per se. For example, "*between 20 and 30 years old*" and "*young*"

Z. Ma: *Fuzzy Database Modeling of Imprecise and Uncertain Engineering Information*,
StudFuzz **195**, 47–58 (2006)
www.springerlink.com

for the attribute *Age* are imprecise and vague values, respectively. In general, vague information is represented by linguistic terms.

Uncertainty is related to the degree of truth of its attribute value, and it means that one can apportion some, but not all, of our belief to a given value or a group of values. For example, the sentence that *"I am 95 percent sure that Tom is married"* represents information uncertainty. Notice that the random uncertainty is described using probability theory.

The ambiguity means that some elements of the model lack complete semantics, leading to several possible interpretations. For example, it may not be clear if one person's salaries stated are per week or month.

Generally, several different kinds of imperfect information can co-exist with respect to the same piece of information. For example, that it is almost sure that John is very young involves information uncertainty and vagueness simultaneously.

3.2.1 Null and Partial Values

Imprecise values generally denote a set of values in the form of [ai1, ai2, ..., aim] or [ai1, ai2] for the discrete and continuous universe of discourse, respectively, meaning that exactly one of the values is the true value for the single-valued attribute, or at least one of the values is the true value for the multivalued attribute. Therefore, imprecise information has two interpretations, which are disjunctive and conjunctive information.

One kind of imprecise information that has been studied extensively is the well-known null values (Codd, 1986 & 1987; Motor, 1990; Parsons, 1996; Zaniola, 1986), which were originally called incomplete information. The possible interpretations of null values include

- *"existing but unknown"* (denoted by *unk* or φ in this thesis),
- *"nonexisting"* or "inapplicable" (denoted by *inap* or $\perp$), and
- *"no information"* (denoted by *nin*).

A null value on a multivalued object, however, means an "open null value" (denoted by *onul*) (Gottlob and Zicari, 1988), i.e., the value may not exist, has exactly one unknown value, or has several unknown values.

Null values with the semantics of "existent but unknown" can be considered as the special type of partial values that the true value can be any one value in the corresponding domain, i.e., an applicable null value corresponds to the whole domain. The notion of a partial value is illustrated as follows (DeMichiel, 1989; Grant, 1979).

Definition. A partial value on a universe of discourse U corresponds to a finite set of possible values in which exactly one of the values in the set is the true value, denoted by $\{a_1, a_2, ..., a_m\}$ for discrete U or $[a_1, a_n]$ for continua U, in which $\{a_1, a_2, ..., a_m\} \subseteq U$ or $[a_1, a_n] \subseteq U$. Let η be a partial value, then *sub* (η) and *sup* (η) are used to represent the minimum and maximum in the set.

Note that crisp data can also be viewed as special cases of partial values. A crisp data on discrete universe of discourse can be represented in the form of $\{p\}$, and a crisp data on continua universe of discourse can be represented in the form of $[p, p]$. Moreover, a partial value without containing any element is called an *empty partial value*, denoted by $\perp$. In fact, the symbol $\perp$ means an inapplicable missing data (Codd, 1986 & 1987). Null values, partial values, and crisp values are thus represented with a uniform format.

3.2.2 Probabilistic Values

Information with a stochastic nature is very common in real-world applications. In order to represent such random uncertainty, probabilistic values are used (Barbara *et al.*, 1992; Cavallo and Pittarelli, 1987; Dey and Sarkar, 1996; Eiter *et al.*, 2001; Lakshmanan *et al.*, 1997; Pittarelli, 1994; Zimanyi, 1997). For a discrete universe of discourse U, a probabilistic value in U is described by a probability distribution ψ_P. Here, a probabilistic measure Prob (u) for each $u \in U$ is needed, which denotes the probability that ψ_P takes u, where $0 \le \text{Prob}(u) \le 1$. Formally, the probability distribution ψ_P is represented as follows.

$$\psi_P = \{\text{Prob}(u_1)/u_1, \text{Prob}(u_2)/u_2, ..., \text{Prob}(u_n)/u_n\}$$

It should be noted that, for a probability distribution ψ_P, the following must hold.

$$\sum_{i=1}^{n} \text{Prob}(u_i) \le 1$$

3.2.3 Fuzzy Values

A large number of data models have been proposed to handle uncertainty and vagueness. Most of these models are based on the same paradigms. Vagueness and uncertainty are generally modeled with fuzzy sets and possibility theory (Zadeh, 1965 & 1978) and any of the existing approaches dealing with imprecision and uncertainty are based on the theory of fuzzy sets and possibility theory.

Fuzzy set theory (Zadeh, 1965) is a generalization of the crisp sets. In a crisp set, an element may or may not be included in the crisp set. In other words, an element may or may not be a member of the set. There are only two situations. But it is possible that an element may have partial membership with regard to a set. In order to represent the set that the elements are included in the set uncertainly, the concept of fuzzy sets was introduced by Lofti Zadeh in 1965 (Zadeh, 1965). Since then fuzzy sets have been infiltrating into almost all branches of pure and applied mathematics that are set-theory-based and have been applied in a number of real applications crossing over a broad realm of domains and disciplines (Cox, 1995; Munakata and Jani, 1994).

For a universe of discourse U, a fuzzy value in U is described by a fuzzy set F_μ. Here, a membership degree $\mu_F(u)$ for each $u \in U$ is needed, which denotes the possibility that u is the element of the fuzzy set, where $0 \leq \mu_F(u) \leq 1$. Formally, the fuzzy set F_μ is represented as follows.

$$F_\mu = \{\mu_F(u_1)/u_1, \mu_F(u_2)/u_2, ..., \mu_F(u_n)/u_n\}$$

Not being the same as probability distribution ψ_P, it is possible that the sum of the membership degrees of all elements in F_μ is greater than 1.

3.3 Representations of Fuzzy Sets and Possibility Distributions

Fuzzy data was originally described as a fuzzy set by Zadeh (1965). Let U be a universe of discourse. A fuzzy value on U is characterized by a fuzzy set F in U. A membership function

$\mu_F: U \rightarrow [0, 1]$

is defined for the fuzzy set F, where $\mu_F(u)$, for each $u \in U$, denotes the degree of membership of u in the fuzzy set F. Thus the fuzzy set F is described as follows:

$$F = \{\mu_F(u_1)/u_1, \mu_F(u_2)/u_2, ..., \mu_F(u_n)/u_n\}$$

When U is an infinite set, then the fuzzy set F can be represented by

$$F = \int_{u \in U} \mu_F(u)/u$$

Definition: A fuzzy set F of the universe of discourse U is convex if and only if for all u_1, u_2 in U,

$$\mu_F \left(\lambda u_1 + (1 - \lambda)\, u_2\right) \geq \min\left(\mu_F\left(u_1\right), \mu_F\left(u_1\right)\right),$$

where $\lambda \in [0, 1]$.

Definition: A fuzzy set F of the universe of discourse U is called a normal fuzzy set if $\exists\, u \in U, \mu_F(u) = 1$.

Definition: A fuzzy number is a fuzzy subset in the universe of discourse U that is both convex and normal.

Now several notions related to fuzzy numbers are discussed. Let U be a universe of discourse and F a fuzzy number in U with the membership function $\mu_F\colon U \to [0,1]$. We have then the following notions:

Support. The set of elements that have non-zero degrees of membership in F is called the support of F, denoted by

$$\operatorname{supp}(F) = \{u \mid u \in U \text{ and } \mu_F(u) > 0\}.$$

Kernel. The set of elements that completely belong to F is called the kernel of F, denoted by

$$\ker(F) = \{u \mid u \in U \text{ and } \mu_F(u) = 1\}.$$

α-Cut. The set of elements whose degrees of membership in F are greater than (greater than or equal to) α, where $0 \leq \alpha < 1$ ($0 < \alpha \leq 1$), is called the strong (weak) α-cut of F, respectively denoted by

$$F_{\alpha+} = \{u \mid u \in U \text{ and } \mu_F(u) > \alpha\}$$

and

$$F_{\alpha} = \{u \mid u \in U \text{ and } \mu_F(u) \geq \alpha\}$$

When the membership function $\mu_F(u)$ above is explained to be a measure of the possibility that a variable X has the value u, where X takes values in U, a fuzzy value is described by a possibility distribution π_X (Zadeh, 1978).

$$\pi_X = \{\pi_X(u_1)/u_1,\ \pi_X(u_2)/u_2,\ \ldots,\ \pi_X(u_n)/u_n\}$$

Here, $\pi_X(u_i)$, $u_i \in U$ denotes the possibility that u_i is true. Let π_X and F be the possibility distribution representation and the fuzzy set representation for a fuzzy value, respectively. It is clear that $\pi_X = F$ is true (Raju and Majumdar, 1988).

3.4 Operations on Fuzzy Sets

In the following, several operations on fuzzy sets are defined to manipulate fuzzy sets (as well as possibility distributions), which include *set operations, arithmetic operations, relational operations,* and *logical operations.*

3.4.1 Set Operations

Let A and B be fuzzy sets in the same universe of discourse U with the membership functions μ_A and μ_B, respectively. Then we have the following:

Union. The union of fuzzy sets A and B, denoted $A \cup B$, is a fuzzy set on U with the membership function $\mu_{A \cup B} \colon U \to [0, 1]$, where

$$\forall u \in U, \mu_{A \cup B}(u) = \max(\mu_A(u), \mu_B(u)).$$

Intersection. The intersection of fuzzy sets A and B, denoted $A \cap B$, is a fuzzy set on U with the membership function $\mu_{A \cap B} \colon U \to [0, 1]$, where

$$\forall u \in U, \mu_{A \cap B}(u) = \min(\mu_A(u), \mu_B(u)).$$

Complementation. The complementation of fuzzy set $\bar{A}$, denoted by $\bar{A}$, is a fuzzy set on U with the membership function $\mu_{\bar{A}} \colon U \to [0, 1]$, where

$$\forall u \in U, \mu_{\bar{A}}(u) = 1 - \mu_A(u).$$

The α-cut of a fuzzy number corresponds to an interval. Let A and B be the fuzzy numbers of the universe of discourse U and let A_α and B_α be the α-cuts of the fuzzy numbers A and B, respectively, where
$A_\alpha = [x_1, y_1]$ and $B_\alpha = [x_2, y_2]$.
Then we have

$$(A \cup B)_\alpha = A_\alpha \underline{\cup} B_\alpha \text{ and } (A \cap B)_\alpha = A_\alpha \underline{\cap} B_\alpha,$$

where $\underline{\cup}$ and $\underline{\cap}$ denote the union operator and intersection operator between two intervals, respectively. The $A_\alpha \underline{\cup} B_\alpha$ and $A_\alpha \underline{\cap} B_\alpha$ are defined as follows:

$$A_\alpha \underline{\cup} B_\alpha = \begin{cases} [x_1, y_1] \ or \ [x_2, y_2], \text{if } A_\alpha \cap B_\alpha = \Phi \\ [\min(x_1, x_2), \max(y_1, y_2)], \text{if } A_\alpha \cap B_\alpha \neq \Phi \end{cases}$$

$$A_\alpha \underline{\cap} B_\alpha = \begin{cases} \Phi, \text{if } A_\alpha \cap B_\alpha = \Phi \\ [\max(x_1, x_2), \min(y_1, y_2)], \text{if } A_\alpha \cap B_\alpha \neq \Phi \end{cases}$$

The operations on fuzzy sets satisfy the properties that are satisfied by the operations defined for the classical sets. Let A, B, and C be fuzzy sets in U. Then

- Commutativity laws: $A \cup B = B \cup A$, $A \cap B = B \cap A$,
- Associativity laws: $(A \cup B) \cup C = A \cup (B \cup C)$, $(A \cap B) \cap C = A \cap (B \cap C)$,
- Distribution laws: $A \cup (B \cap C) = (A \cup B) \cap (A \cup C)$, $A \cap (B \cup C) = (A \cap B) \cup (A \cap C)$,
- Absorption laws: $A \cup (A \cap B) = A$, $A \cap (A \cup B) = A$,
- Idempotency laws: $A \cup A = A$, $A \cap A = A$, and
- de Morgan laws: $\overline{A \cup B} = \overline{A} \cap \overline{B}$, $\overline{A \cap B} = \overline{A} \cup \overline{B}$.

3.4.2 Arithmetic Operations

The arithmetic operations on fuzzy sets can be defined by using Zadeh's extension principle, which can also be referred to maximum-minimum principle sometimes. Let X_1, X_2, ..., X_n and Y be ordinary sets, f be a mapping from $X_1 \times X_2 \times \ldots \times X_n$ to Y such that $y = f(x_1, x_2, \ldots, x_n)$, P (X_i) and P (Y) be the power sets of X_i and Y ($0 \le i \le n$), respectively. Here, P $(X_i) = \{C | C \subseteq X_i\}$ and P (Y) = $\{D | D \subseteq Y\}$. Then f induces a mapping from P $(X_1) \times$ P $(X_2) \times \ldots \times$ P (X_n) to P (Y) with

$$f(C_1, C_2, \ldots, C_n) = \{f(x_1, x_2, \ldots, x_n) | x_i \in C_i, 0 \le i \le n\},$$

where $C_i \subseteq X_i$, $0 \le i \le n$. Now, let F (X_i) be the class of all fuzzy sets on X_i, i.e., F $(X_i) = \{\}$, $0 \le i \le n$ and F (Y) be the class of all fuzzy sets on Y, i.e., F (Y) = $\{\}$, then f induces a mapping from F $(X_1) \times$ F $(X_2) \times \ldots \times$ F (X_n) to F (Y) such that for all Ai $\in$ F (X_i), $f(A_1, A_2, \ldots, A_n)$ is a fuzzy set on Y with

$$f(A_1, A_2, \ldots, A_n)(y) =$$
$$\begin{cases} \sup_{\substack{f(x1,x2,\ldots,xn)=y \\ xi \in Xi\,(i=1,2,\ldots,n)}} (\min(\mu_{A1}(x1), \mu_{A2}(x2), \ldots, \mu_{An}(xn)), f^{-1}(y)) \ne \Phi \\ 0, f^{-1}(y) = \Phi \end{cases}$$

Let A and B be fuzzy sets on the same universe of discourse U with the membership functions μ_A and μ_B, respectively, and "θ" be an infix operator. $A\,\theta\,B$ is a fuzzy set on U with the membership function $\mu_{A\,\theta\,B}: U \to [0, 1]$, where

$$\forall z \in U, \mu_{A \theta B}(z) = \max_{z = x \theta y}(\min(\mu_A(x), \mu_B(y))).$$

3.4.3 Relational Operations

The classical relational operations are various kinds of comparison operations on the classical sets, including *equal to* ($=$), *not equal to* ($\neq$), *greater than* ($>$), *greater than or equal to* ($\geq$), *less than* ($<$), and *less than or equal to* ($\leq$). The definitions of the relational operations of fuzzy sets are essentially related to the closeness measures between fuzzy sets and the given thresholds. So, we use the notion of semantic equivalence degree to define the relational operations on fuzzy sets.

Some approaches for assessing fuzzy data redundancy have been proposed in literature to measure the semantic relationship between fuzzy data. But it was shown in (Ma, Zhang and Ma, 1999) that there are some inconsistencies when the proposed approaches are applied to assess the redundancy of fuzzy data represented by the possibility distributions. In addition, the semantic measures of fuzzy data were mainly focused on equivalence relationship. To eliminate counterintuitive results in the proposed approaches and to assess the semantic relationship of fuzzy data, including equivalence and inclusion, we use the notions of semantic space and semantic inclusion degree.

Not being the same as a precise value, a fuzzy value is a set of points in space instead of a point in the universe of discourse. The semantics of a fuzzy value expressed by possibility distribution corresponds to an area in space, the so-called semantic space.

Definition: For fuzzy data, its semantics correspond to an area in space, where the universe of discourse is its X-axis and possibility is its Y-axis.

The semantic relationship between two fuzzy data can be described by the relationship between their semantic spaces. Let SS (A) and SS (B) be semantic spaces of fuzzy data A and B, respectively. If SS (A) $\supseteq$ SS (B) (or SS (A) $\subseteq$ SS (B)), A semantically includes B (or A is semantically included by B). If SS (A) $\supseteq$ SS (B) and SS (A) $\subseteq$ SS (B), A and B are semantically equivalent to each other. It is clear that semantic equivalence is a particular case of semantic inclusion. We can employ semantic inclusion degree to measure the semantic relationship of fuzzy data.

Definition: Let A and B be two fuzzy data, and their semantic spaces be SS (A) and SS (B), respectively. Let SID (A, B) denotes the degree that A semantically includes B. Then

$$\text{SID}(A, B) = (\text{SS}(B) \cap \text{SS}(A))/\text{SS}(B)$$

For two fuzzy data A and B, the meaning of SID (A, B) is the percentage of the semantic space of B which is wholly included in the semantic space of A. Following this definition, the concept of equivalence degree can be easily drawn as follows.

Definition: Let A and B be two fuzzy data and SID (A, B) be the degree that A semantically includes B. Let SE (A, B) denote the degree that A and B are equivalent to each other. Then

$$SE\ (A, B) = \min\ (SID\ (A, B),\ SID\ (B, A))$$

For two fuzzy sets represented the possibility distributions, the semantic inclusion degree is defined as follows.

Definition: Let $U = \{u_1, u_2, \ldots, u_n\}$ be the universe of discourse. Let π_A and π_B be two fuzzy data on U based on possibility distribution. Let $\pi_X\ (u_i)$, $u_i \in U$, denotes the possibility that u_i is true. The SID (π_A, π_B) is defined as

$$SID\ (\pi_A, \pi_B) = (SS\ (B) \cap SS\ (A))/SS\ (B) =$$
$$\sum_{i=1}^{n} \min_{u_i, u_j \in D}\ (\pi_B\ (u_i), \pi_A\ (u_i))\ /\ \sum_{i=1}^{n} \pi_B\ (u_i)$$

Based on the semantic equivalence degree for fuzzy sets, we can define the relational operations on fuzzy sets. Let A and B be fuzzy sets on the same universe of discourse U with the membership functions μ_A and μ_B, respectively, and β be a given threshold value. Then

- $A \approx_\beta B$ if $SE\ (A, B) \geq \beta$,
- $A \not\approx_\beta B$ if $SE\ (A, B) < \beta$,
- $A \succ_\beta B$ if $SE\ (A, B) < \beta$ and $\max\ (\mathrm{supp}\ (A)) > \max\ (\mathrm{supp}\ (B))$,
- $A \prec_\beta B$ if $SE\ (A, B) < \beta$ and $\max\ (\mathrm{supp}\ (A)) < \max\ (\mathrm{supp}\ (B))$,
- $A \succeq_\beta B$ if $A \succ_\beta B$ or $A \approx_\beta B$, and
- $A \preceq_\beta B$ if $A \prec_\beta B$ or $A \approx_\beta B$.

3.4.4 Logical Operations

Three logical operations *fuzzy not* ($\tilde{\backsim}$), *fuzzy and* ($\tilde{\wedge}$), and *fuzzy or* ($\tilde{\vee}$), which operands are fuzzy Boolean value(s) represented by fuzzy sets, are defined in this section. Fuzzy logical operations hereby depend on the representation of fuzzy Boolean values and fuzzy logic.

Fuzzy and ($\tilde{\wedge}$). *Fuzzy and* can be defined with "intersection" kinds of operations such as "min" operation and its result is a fuzzy Boolean value. Let A: $\mu_A\ (u)$ and B: $\mu_B\ (u)$ be two fuzzy Boolean values represented by fuzzy sets on the same universe of discourse U. Then

$$A \; \tilde{\wedge} \; B: \min \; (\mu_A \, (u), \; \mu_B \, (u)), \; u \in U.$$

Fuzzy or ($\tilde{\vee}$). *Fuzzy or* can be defined with "union" kinds of operations such as "max" operation and its result is a fuzzy Boolean value. Let A: μ_A (u) and B: μ_B (u) be two fuzzy Boolean values represented by fuzzy sets on the same universe of discourse U. Then

$$A \; \tilde{\vee} \; B: \max \; (\mu_A \, (u), \; \mu_B \, (u)), \; u \in U.$$

Fuzzy not ($\tilde{\backsimeq}$). *Fuzzy not* can be defined with "complementation" kinds of operations such as "subtraction" operation and its result is a fuzzy Boolean value. Let A: μ_A (u) be a fuzzy Boolean values represented by fuzzy sets on the universe of discourse U. Then

$$\tilde{\backsimeq} \; A: (1 - \mu_A \, (u)), \; u \in U.$$

3.5 Summary

In this chapter, we have presented certain very basic definition and results on fuzzy sets and related topics. There are many more mathematical generalizations of fuzzy sets such as vague sets (Gau and Buehrer, 1993) and type-2 fuzzy sets (Mizumoto and Tanaka, 1976; de Tré and de Caluwe, 2003). From the applications point of view, the notion of type-2 fuzzy sets is rather interesting. The type-2 fuzzy sets have recently applied to areas like artificial intelligence (AI), forecasting of time series, knowledge-mining, and digital communications etc.

Viewed from imprecise and uncertain information modeling, the fuzzy relational databases have been extensively studied in last two decades and current efforts have been concentrated on the fuzzy object-oriented databases and the fuzzy conceptual data modeling.

References

Aliev, R. A., Aliev, R. R. and Aliev, R. (2001), Soft Computing and Its Applications, World Scientific Publishing Company.

Barbara, D., Garcia-Molina, H. and Porter, D. (1992), The management of probabilistic data, IEEE Transactions on Knowledge and Data Engineering, 4 (5): 487-502.

Bosc, P. and Prade, H. (1993), An introduction to fuzzy set and possibility theory based approaches to the treatment of uncertainty and imprecision in database management systems", Proceedings of the Second Workshop on Uncertainty Management in Information Systems: From Needs to Solutions.

Cavallo, R. and Pittarelli, M. (1987), The theory of probabilistic databases, Proceedings of the 13th VLDB Conference, 71-81.

Codd, E. F. (1986), Missing information (applicable and inapplicable) in relational databases, SIGMOD Record, 15: 53-78.

Codd, E. F. (1987), More commentary on missing information in relational databases (applicable and inapplicable information), SIGMOD Record, 16 (1): 42-50.

Cox, E. (1995), Fuzzy Logic for Business and Industry, Massachusetts: Charles River Media, Inc.

de Tré, G. and de Caluwe, R. (2003), Level-2 fuzzy sets and their usefulness in object-oriented database modelling, Fuzzy Sets and Systems, 140 (1): 29-49.

DeMichiel, L. G. (1989), Resolving database incompatibility: an approach to performing relational operations over mismatched domains, IEEE Transactions on Knowledge and Data Engineering, 1 (4): 485-493.

Dey, D. and Sarkar, S. A. (1996), Probabilistic relational model and algebra, ACM Transactions on Database Systems, 21 (3): 339-369.

Eiter, T., Lu, J. J., Lukasiewicz, T. and Subrahmanian, V. S. (2001), Probabilistic object bases, ACM Transactions on Databases Systems, 26 (3): 264-312.

Gottlob, G. and Zicari, R. (1988), Closed world databases opened through null values, Proceedings of the 1988 International Conference on Very Large Data Bases, 50-61.

Grant, J., 1979, Partial values in a tabular database model, Information Processing Letters, 9 (2): 97-99.

Gau, W. L. and Buehrer, D. J. (1993), Vague sets, IEEE Transactions on Systems, Man, and Cybernetics, 23 (2): 610-614.

Lakshmanan, L. V. S., Leone, N., Ross, R. and Subrahmanian, V. S. (1997), ProbView: A flexible probabilistic database system, ACM Transactions on Database Systems, 22 (3): 419-469.

Mizumoto, M. and Tanaka, K. (1976), Some properties of fuzzy sets of type 2, Information and Control, 31 (4): 312-340.

Motor, A. and Smets, P., 1997, Uncertainty Management in Information Systems: From Needs to Solutions, Kluwer Academic Publishers.

Motor, A. (1990), Accommodation imprecision in database systems: issues and solutions, ACM SIGMOD Record, 19 (4): 69-74.

Munakata, T. and Jani, Y. (1994), Fuzzy systems: an overview, Communications of The ACM, 37 (3): 69-76.

Parsons, S. (1996), Current approaches to handling imperfect information in data and knowledge bases, IEEE Transactions on Knowledge Data Engineering, 8: 353–372.

Pittarelli, M. (1994), An algebra for probabilistic databases, IEEE Transactions on Knowledge and Data Engineering, 6 (2): 293-303.

Raju, K. V. S. V. N. and Majumdar, A. K. (1988), Fuzzy functional dependencies and lossless join decomposition of fuzzy relational database system, ACM Transactions on Database Systems, 13(2): 129-166.

Zadeh, L. A. (1965), Fuzzy sets, Information and Control, 8 (3): 338-353.

Zadeh, L. A. (1975), The concept of a linguistic variable and its application to approximate reasoning, Information Sciences, 8: 119-249 & 301-357; 9: 43-80.

Zadeh, L. A. (1978), Fuzzy sets as a basis for a theory of possibility, Fuzzy Sets and Systems, 1 (1): 3-28.

Zaniolo, C. (1984), Database systems with null values, Journal of Computer and System Sciences, 28 (2): 142- 166.

Zimanyi, E. (1997), Query evaluation in probabilistic relational databases, Theoretical Computer Science, 171: 179-219.

4 The Fuzzy ER/EER and UML Data Models

4.1 Introduction

Information imprecision and uncertainty exist in real-world applications. It is especially true in some non-traditional applications (e.g., decision-making and expert systems). Currently the fuzzy set theory has extensively been applied for representing imprecise and uncertain information. As described in Chapter 1, the conceptual (semantic) data models for conceptual data modeling provide the designers with powerful mechanisms in generating the most complete specification from the real world. The conceptual data models, e.g., ER/EER and UML, represent both complex structures of entities and complex relationships among entities as well as their attributes. So the conceptual data models play an important role in conceptual data modeling and database conceptual design. In order to deal with complex objects and imprecise and uncertain information in conceptual data modeling, one needs fuzzy extension to conceptual data models, which allow imprecise and uncertain information to be represented and manipulated at a conceptual level.

It will be shown in Chapter 7 that fuzzy databases have been extensively studied in last two decades in the context of the relational database model and current efforts have been concentrated on the fuzzy object-oriented databases. But less research has been done in the fuzzy conceptual data modeling. The fuzzy set theory was first applied to some of the basic ER concepts in (Zvieli and Chen, 1986). Fuzzy entity sets, fuzzy relationship sets and fuzzy attribute sets were introduced in addition to fuzziness in entity and relationship occurrences and in attribute values. Consequently, fuzzy extension to the ER algebra (Chen, 1976) has been sketched. Other efforts to extend the ER model can be found in (Ruspini, 1986; Vandenberghe, 1991; Chaudhry *et al.*, 1999; Vert *et al.*, 2000). Chaudhry *et al.* (1999) proposed a method for designing fuzzy relational databases (FRDBs) following the extension of the ER model of (Zvieli and Chen, 1986) taking special interest in converting crisp databases into fuzzy ones. In (Ruspini,

Z. Ma: *Fuzzy Database Modeling of Imprecise and Uncertain Engineering Information*, StudFuzz **195**, 59–77 (2006)
www.springerlink.com

1986), an extension of the ER model with fuzzy values in the attributes was proposed and a truth value can be associated with each relationship instance. In addition, some special relationships such as *same-object* and *subset-of*, member-of were also introduced. Vandenberghe (1991) applied Zadeh's extension principle to calculate the truth value of propositions. For each proposition, a possibility distribution was defined on the doubleton true, false of the classical truth values. The proposal of Vert *et al.* (2000) was based on the notation used by Oracle and used the fuzzy sets theory to treat data sets as a collection of fuzzy objects, applying the result to the area of geospatial information systems (GISs). Without including graphical representations, the fuzzy extensions of several major EER concepts (superclass/subclass, generalization/specialization, category, and the subclass with multiple superclasses) were introduced in (Chen and Kerre, 1998). More recently, Galindo *et al.* (2004) extended the EER models by relaxing some constraints with fuzzy quantifiers.

In this chapter, multiple granularity of information fuzziness in ER/EER and UML data models will be identified. Based on possibility distribution theory, the fuzzy extension to these data models and the graphical representations will be presented.

4.2 The Fuzzy ER/EER Data Models

The extension to the ER data model to incorporate fuzziness was proposed by Zvieli and Chen (1986), where fuzzy entities, attributes and relationships are represented in the graphical model. Consequently, fuzzy extension to Chen' ER algebra has been sketched.

4.2.1 Three Levels of Fuzziness in ER Data Model

The ER data model is described in three levels. The first level is model level, referring to the entity type, relationship type and attribute. The second level is type/instance level, referring to the instances of an entity type or a relationship type. The third level is attribute value level, referring to the attribute values of an entity or relationship instance. The fuzzy extension to the ER data model is hereby carried out in the three levels.

Zvieli and Chen (1986) allowed fuzzy attributes in entities and relationships and three levels of fuzziness were introduced in the ER model.
- At the first level, entity sets, relationships and attributes may be fuzzy. In other words, they may have a membership degree to the ER model.

- The second level is related to the fuzzy occurrences of entities and relationships.
- The third level concerns the fuzzy values of attributes of special entities and relationships.

Formally, let E, R, and A be the fuzzy entity type set, fuzzy relationship type set, and fuzzy attribute set of the fuzzy ER model, respectively, and μ_E, μ_R, and μ_A be their membership functions. Then

- for an entity type, say E_i, we have $\mu_E(E_i)/E_i$, where $\mu_E(E_i)$ is the degree of E_i belonging to E and $0 \le \mu_E(E_i) \le 1$,
- for a relationship type, say R_i, we have $\mu_R(R_i)/R_i$, where $\mu_R(R_i)$ is the degree of R_i belonging to R and $0 \le \mu_R(R_i) \le 1$, and
- for an attribute of entity type or relationship type, say A_i, we have $\mu_A(A_i)/A_i$, where $\mu_A(A_i)$ is the degree of A_i belonging to A and $0 \le \mu_A(A_i) \le 1$.

Figure 4.1 shows fuzzy ER diagram notations with the first level of fuzziness.

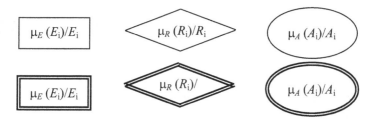

Fig. 4.1. The fuzzy ER diagram notations with the first level of fuzziness

Fuzziness at the second and third levels is represented with symbol "f" labeled in the fuzzy ER diagram in (Zvieli and Chen, 1986).

4.2.2 The Fuzzy EER Data Model

Based on the issues of Zvieli and Chen (1986), this section introduces some notions and notations to model fuzzy information in the EER data model.

Fuzzy Attribute

In a fuzzy EER model, first attributes may have first level of fuzziness in the fuzzy ER model, referring to fuzzy attribute sets. In addition, attribute values may be fuzzy ones based on possibility distribution, which are the same as the third level of fuzziness in the fuzzy ER model. Moreover, there are two kinds of interpretation of a fuzzy attribute value because of single-valued and multivalued attributes, which are a fuzzy disjunctive attribute value and a fuzzy conjunctive attribute value. In connection to these two kinds of fuzzy attribute values, the fuzzy attributes can be classified into the fuzzy disjunctive one and the fuzzy conjunctive one. Also a composite attribute may be fuzzy too, either disjunctive or conjunctive, if one of its components is fuzzy.

The graphical representations of the fuzzy attributes mentioned above are shown in Figure 4.2.

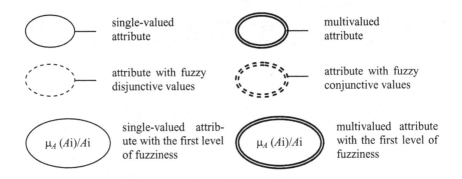

single-valued attribute

multivalued attribute

attribute with fuzzy disjunctive values

attribute with fuzzy conjunctive values

single-valued attribute with the first level of fuzziness

multivalued attribute with the first level of fuzziness

Fig. 4.2. Attributes in the fuzzy EER model

Fuzzy Entity and Relationship

There exist two levels of fuzziness in entity and relationship types of the fuzzy EER model, which are the first level fuzziness and the second level of fuzziness.

To graphically represent the entity type with the first level of fuzziness, one can place membership degrees inside the diagrams of entity and relationship types in the fuzzy EER model. Formally let E_i be an entity type and $\mu\ (E_i)$ be its degree of membership in the model, then "$\mu\ (E_i)/E_i$" is enclosed in the rectangle. If $\mu\ (E_i) = 1$, "$1/E_i$" is usually denoted by "E_i" simply. In a similar way, the relationship type with the first level of fuzziness can be represented.

The graphical representations of the entity and relationship types with the first level of fuzziness are shown in Figure 4.3. The graphical representations of the entity and relationship types with the second level of fuzziness are shown in Figure 4.4.

Fig. 4.3. Entity and relationship types with the first level of fuzziness

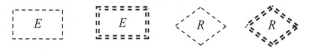

Fig. 4.4. Entity and relationship types with the second level of fuzziness

Note that in the fuzzy EER model, an entity type with the first and second level of fuzziness can be an weak one, a relationship type with the first and second level of fuzziness can be an ownership one, and an attribute of entity type or relationship type can be simple one, multivalued, composite, or fuzzy (disjunctive or conjunctive).

Fuzzy Generalization/Specialization

The subclass-superclass relationship between entity types is described by generalization and specialization in the EER model. An entity type, say S, is called a subclass of another entity type, say E, and meanwhile E is called a superclass of S, if and only for any entity instance of S, it is also the entity instance of E. But there exist three levels of fuzziness in the fuzzy EER model. As we know, the second level of fuzziness is the fuzziness in entity type/instance. An entity instance may have a membership degree with respect to an entity type. The subclass-superclass relationship between such entity types may be fuzzy and we have to redefine it.

Formally let E and S be two entity types on universe of discourse U and they are both fuzzy sets with membership functions μ_E and μ_S, respectively. Then S is a fuzzy subclass of E and E is a fuzzy superclass of S if and only if the following is true.

$$(\forall e)\,(e \in U \wedge \mu_S(e) \le \mu_E(e))$$

Now consider the situation that superclass E has multiple subclasses S_1, S_2, ..., S_n and their membership functions are μ_E, μ_{S1}, μ_{S2}, ..., and μ_{Sn}, respectively. Then

$$(\forall e)\,(e \in U \wedge \max(\mu_{S1}(e), \mu_{S2}(e), ..., \mu_{Sn}(e)) \le \mu_E(e))$$

That is, the degree that an entity instance belongs to any of subclasses is not greater than the degree that the entity instance belongs to superclass of the subclasses.

After defining the subclass-superclass relationship between the entities with the second level of fuzziness, let us focus on the fuzzy generalization and specialization. A generalization defines a superclass from several entity types, generally being with some common features, while a specialization defines several subclasses from an entity type according to a certain feature. Let E be fuzzy superclass of fuzzy subclasses S_1, S_2, ..., S_n and they have membership functions μ_E, μ_{S1}, $\mu_{S2, ...}$, and μ_{Sn}, respectively, Then

$$(\forall\ e)\ (\forall\ S)\ (e \in U \wedge S \in \{S_1, S_2, ..., S_n\} \wedge \mu_S\ (e) \leq \mu_E\ (e)).$$

This means that for each subclass, the degree that an entity instance belongs to it must be less than or equal to the degree that the entity instance belongs to superclass of the subclasses.

In the classical EER model, the total, partial, disjoint and overlapping specializations can be identified. Correspondingly we have the fuzzy total specialization, fuzzy partial specialization, fuzzy disjoint specialization, and fuzzy overlapping specialization in the fuzzy EER model.

- S_1, S_2, ..., S_n are a fuzzy total specialization of E if

$$(\forall\ e)\ (\exists\ S)\ (e \in E \wedge S \in \{S_1, S_2, ..., S_n\} \wedge 0 < \mu_S\ (e) \leq \mu_E\ (e))$$

It can be seen that for any of entity instances belonging to a superclass with a non-zero degree, it must belong to one of its subclasses with a non-zero degree as well. In addition, the later non-zero degree is not greater than the former non-zero degree.

- S_1, S_2, ..., S_n are a fuzzy partial specialization of E if

$$(\exists\ e)\ (\forall S)\ (e \in E \wedge S \in \{S_1, S_2, ..., S_n\} \wedge \mu_E\ (e) > 0 \wedge \mu_S\ (e) = 0)$$

That is, there exists entity instance that belongs to superclass with a non-zero degree but belongs to any of subclasses with a zero degree.

- S_1, S_2, ..., S_n are disjoint if

$$(\nexists e)\ (\forall S)\ (\forall S')\ (e \in U \wedge S \in \{S_1, S_2, ..., S_n\} \wedge S' \in \{S_1, S_2, ..., S_n\} \wedge$$
$$\min\ (\mu_S\ (e),\ \mu_{S'}\ (e)) > 0)$$

That is, there exists no entity instance that belongs to more than one subclass each with a non-zero degree.

- S_1, S_2, ..., S_n are overlapping if

$$(\exists\ e)\ (\forall\ S)\ (\forall\ S')\ (e \in U \wedge S \in \{S_1, S_2, ..., S_n\} \wedge S' \in \{S_1, S_2, ..., S_n\} \wedge$$
$$\min\ (\mu_S\ (e),\ \mu_{S'}\ (e)) > 0)$$

This means that there exists entity instance that belongs to more than one subclass each with a non-zero degree.

Being the inverse process of the fuzzy specialization, the fuzzy generalization can be classified into fuzzy total generalization, fuzzy disjoint generalization, and fuzzy overlapping generalization. Note that there is no fuzzy partial generalization just like the classical EER model.

Fuzzy Category

Let E be a fuzzy category of fuzzy entity sets S_1, S_2, ..., S_n and their membership functions are μ_{Ei}, μ_{S1}, $\mu_{S2, ...}$, and μ_{Sn}, respectively. Then

$$(\forall\ e)\ (\exists\ S)\ (e \in E \land S \in \{S_1, S_2, ..., S_n\} \land \mu_S\ (e) \geq \mu_E\ (e) > 0)$$

This means that for any of entity instances belonging to category with a non-zero degree, it must belong to one of its superclasses with a non-zero degree as well. In addition, the former non-zero degree is not greater than the later non-zero degree.

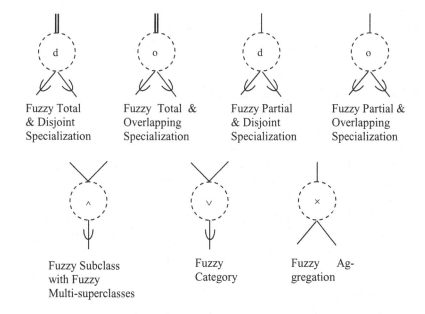

Fuzzy Total & Disjoint Specialization	Fuzzy Total & Overlapping Specialization	Fuzzy Partial & Disjoint Specialization	Fuzzy Partial & Overlapping Specialization

Fuzzy Subclass with Fuzzy Multi-superclasses   Fuzzy Category   Fuzzy Aggregation

Fig. 4.5. Fuzzy EER diagram notations of specialization, category, and aggregation

Note that the fuzzy category is different from the fuzzy subclass with more than one fuzzy superclass. Let E be a fuzzy subclass and S_1, S_2, ..., S_n

be its fuzzy superclasses and their membership functions are μ_E, μ_{S1}, $\mu_{S2, ...}$, and μ_{Sn}, respectively. Then

$$(\forall\ e)\ (\forall\ S)\ (e \in Ei \wedge S \in \{S_1, S_2, ..., S_n\} \wedge \mu_S\ (e) \geq \mu_{Ei}\ (e) > 0)$$

It is clear that they are different each other.

Fuzzy Aggregation

Let E be a fuzzy aggregation of fuzzy entity sets S_1, S_2, ..., and S_n and their membership functions are μ_{Ei}, μ_{S1}, $\mu_{S2, ...}$, and μ_{Sn}, respectively. Then

$$(\forall\ e)\ (\exists\ e_1)\ (\exists\ e_2)\ ...\ (\exists\ e_n)\ (e \in E \wedge e_1 \in S_1 \wedge e_2 \in S_2 \wedge\ ...\ \wedge e_n \in S_n$$
$$\wedge \mu_E\ (e) = \mu_{S1}\ (e_1) \times \mu_{S2}\ (e_2) \times\ ...\ \times \mu_{Sn}\ (e_n) \neq 0).$$

That is, for any of entity instances belonging to aggregation with a non-zero degree, it can be broken down into some parts and as an entity instance, each part must belong to one of its part entity types with a non-zero degree. In addition, the product of all of the later non-zero degrees forms the former non-zero degree.

The diagrammatic notations of the fuzzy specialization, category and aggregation in the fuzzy EER model are shown in Figure 4.5.

4.3 The Fuzzy UML Data Model

UML provides a collection of models to capture the many aspects of a software system (Booch *et al.*, 1998; OMG, 2003). Notice that while the UML reflects some of the best OO modeling experiences available, it suffers from some lacks of necessary semantics. One of the lacks can be generalized as the need to handle imprecise and uncertain information although such information exist in knowledge engineering and databases and have extensively being studied (Parsons, 1996). In the following, we propose the fuzzy UML data model.

4.3.1 Fuzzy Class

The objects having the same properties are gathered into classes that are organized into hierarchies. Theoretically, a class can be considered from two different viewpoints:
- an extensional class, where the class is defined by the list of its object instances (objects for short), and

- an intensional class, where the class is defined by a set of attributes and their admissible values.
  A class may be fuzzy because of the following several reasons.
- Some objects with similar properties are fuzzy ones. A class defined by these objects may be fuzzy. Then the objects belong to the class with membership degree of [0, 1].
- When a class is intensionally defined, the domains of some attributes may be fuzzy and a fuzzy class is formed.
- The subclass produced by a fuzzy class by means of specialization and the superclass produced by some classes (in which there is at least one class who is fuzzy) by means of generalization are also fuzzy.

Following on the footsteps of (Zvieli and Chen, 1986), we introduce three levels of fuzziness into the UML classes. These three levels of fuzziness are defined as follows:

- At the first level, classes and attribute sets of class may be fuzzy, i.e., they have a membership degree to the model.
- The second level is related to the fuzzy occurrences of objects.
- The third level concerns the fuzzy values of attributes of special objects.

In order to model the first level of fuzziness, i.e., an attribute or a class with degree of membership, the attribute or class name should be followed by a pair of words WITH *mem* DEGREE, where $0 \leq mem \leq 1$ is used to indicate the degree that the attribute belongs to the class or the class belongs to the data model (Marín *et al.*, 2000). Assume, for example, that we have a class "Employee" with attribute "Office" in the data model. Then "Employee *WITH 0.8 DEGREE*" and "Office *WITH 0.6 DEGREE*" are the class and attribute with the first level of fuzziness, respectively. That means class "Employee" belongs to the data model with 0.8 degree while attribute "Office" belongs to the class with 0.6 degree. Generally, an attribute or a class will not be declared when its degree is 0. In addition, "WITH 1.0 DEGREE" can be omitted when the degree of an attribute or a class is 1. Also note that attribute values may be fuzzy. In order to model the third level of fuzziness, a keyword FUZZY is introduced and is placed in the front of the attribute. As to the second level of fuzziness, we must indicate the degree of membership that an object of the class belongs to the class. To this purpose, an additional attribute is introduced into the class to represent object membership degree to the class, which attribute domain is [0, 1]. We denote such special attribute with μ in the paper. In order to differentiate the class with the second level of fuzziness, we use a dashed-outline rectangle to denote such a class.

Figure 4.6 shows a fuzzy class *Student*. Here, attribute *Age* may take fuzzy values, namely, its domain is fuzzy. The students may or may not have their offices. It is not known for sure if *Student* has attribute *Office*. But we know the students may have their offices with low possibility, say 0.4. So attribute *Office* uncertainly belongs to the class *Students*. This class has the fuzziness at the first level and we use "with 0.4 membership degree" to describe the fuzziness in the class definition. In addition, we may not determine if an object is the instance of the class because the class is fuzzy. So an additional attribute μ is introduced into the class.

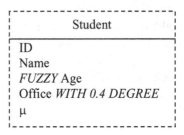

Fig. 4.6. A fuzzy class in the fuzzy UML

4.3.2 Fuzzy Generalization

The concept of subclassing is one of the basic building blocks of the object model. A new class, called subclass, is produced from another class, called superclass by means of inheriting all attributes and methods of the superclass, overriding some attributes and methods of the superclass, and defining some new attributes and methods. Since a subclass is the specialization of the superclass, any one object belonging to the subclass must belong to the superclass. This characteristic can be used to determine if two classes have subclass/superclass relationship.

However a class produced from a fuzzy class must be fuzzy. If the former is still called subclass and the later superclass, the subclass/superclass relationship is fuzzy. In other words, a class is a subclass of another class with membership degree of [0, 1] at this moment. Correspondingly, we have the following method for determining subclass/superclass relationship.

- for any one (fuzzy) object, if the membership degree that it belongs to the subclass is less than or equal to the membership degree that it belongs to the superclass, and

- the membership degree that it belongs to the subclass is greater than or equal to the given threshold.

The subclass is then a subclass of the superclass with the membership degree, which is the minimum in the membership degrees to which these objects belong to the subclass.

Definition. Let A and B be (fuzzy) classes and β be a given threshold. We say B is a subclass of A if

$$(\forall e)\, (\beta \leq \mu_B\, (e) \leq \mu_A\, (e)).$$

The membership degree that B is a subclass of A should be $\min_{\mu_B\,(e)\,\geq\,\beta}$ $(\mu_B\,(e))$. Here, e is object instance of A and B in the universe of discourse, and $\mu_A\,(e)$ and $\mu_B\,(e)$ are membership degrees of e to A and B, respectively.

It should be noted that, however, in the above-mentioned fuzzy generalization relationship, we assume that classes A and B can only have the second level of fuzziness. It is possible that classes A or B are the classes with membership degree, namely, with the first level of fuzziness.

Definition. Let two classes A and B be A WITH *degree_A* DEGREE and B WITH *degree_B* DEGREE. Then B is a subclass of A if

$$(\forall e)\, (\beta \leq \mu_B\, (e) \leq \mu_A\, (e)) \wedge ((\beta \leq degree\_B \leq degree\_A).$$

That means that B is a subclass of A only if, in addition to the condition that the membership degrees of all objects to A and B must be greater than or equal to the given threshold and the membership degree of any object to A must be greater than or equal to the membership degree of this object to B, the membership degrees of A and B must be greater than or equal to the given threshold and the membership degree of A must be greater than or equal to the membership degree of B.

Considering a fuzzy superclass A and its fuzzy subclasses B_1, B_2, ..., B_n with object membership degrees μ_A, μ_{B1}, $\mu_{B2, ...}$, and μ_{Bn}, respectively, which may have the degrees of membership *degree_A*, *degree_B$_1$*, *degree_B$_2$*, ..., and *degree_B$_n$*, respectively. Then the following relationship is true.

$$(\forall e)\, (\max\, (\mu_{B1}\, (e),\, \mu_{B2}\, (e),\, ...,\, \mu_{Bn}\, (e)) \leq \mu_A\, (e)) \wedge (\max\, (degree\_B_1,\, degree\_B_2,\, ...,\, degree\_B_n) \leq degree\_A)$$

It can be seen that we can assess fuzzy subclass/superclass relationships by utilizing the inclusion degree of objects to the class. Clearly such assessment is based on the extensional viewpoint of class. When classes are defined with the intensional viewpoint, there is no any object available. Therefore, the method given above cannot be used. At this point, we can use the inclusion degree of a class with respect to another class to determine the relationships between fuzzy subclass and superclass. The basic idea is that since any one object belonging to the subclass should belong to the superclass, the common attribute domains of the superclass should include the common attribute domains of the subclass.

Definition. Let A and B be (fuzzy) classes and the degree that B is the subclass of A be denoted by $\mu (A, B)$. For a given threshold β, we say B is a subclass of A if

$$\mu (A, B) \geq \beta.$$

Here $\mu (A, B)$ is used to calculate the inclusion degree of B with respect to A according to the inclusion degree of the attribute domains of B with respect to the attribute domains of A as well as the weight of attributes.

To figure out or estimate the inclusion degree of two classes, ones need to know the (fuzzy) attribute domains of these two classes and the weight of these attributes. The problem of evaluating the inclusion degree can be found in Chapter 7, where the methods for evaluating the inclusion degree of fuzzy attribute domains and further evaluating the inclusion degree of a subclass with respect to the superclass are discussed in details.

Now let us consider the situation that classes A or B are the classes with membership degree, namely, with the first level of fuzziness.

Definition. Let two classes A and B be A WITH *degree_A* DEGREE and B WITH *degree_B* DEGREE. Then B is a subclass of A if

$$(\mu (A, B) \geq \beta) \wedge ((\beta \leq degree\_B \leq degree\_A).$$

That means that B is a subclass of A only if, in addition to the condition that the inclusion degree of A with respect to B must be greater than or equal to the given threshold, the membership degrees of A and B must be greater than or equal to the given threshold and the membership degree of A must be great than or equal to the membership degree of B.

In subclass-superclass hierarchies, a critical issue is multiple inheritance of class. Ambiguity arises when more than one of the superclasses have common attributes and the subclass does not declare explicitly the class from which the attribute is inherited. At this moment, which conflicting attribute in the superclasses is inherited by the subclass dependents on their weights to the corresponding superclasses (Liu and Song, 2001). Also it should be noticed that in fuzzy multiple inheritance hierarchy, the subclass has different degrees with respect to different superclasses, not being the same as the situation in classical object-oriented databases.

In order to represent a fuzzy generalization relation, a dashed peculiar triangular arrowhead is applied. Figure 4.7 shows a fuzzy generalization relationship. Here classes *Young Student* and *Young Faculty* are all classes with the second level of fuzziness. That means that the classes may have some objects, which belong to the classes with membership degree. These two classes can be generalized into class *Youth*, a class with the second level of fuzziness.

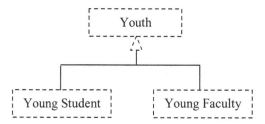

Fig. 4.7. A fuzzy generalization relationship in the fuzzy UML

4.3.3 Fuzzy Aggregation

An aggregation captures a whole-part relationship between an aggregate and a constituent part and these constituent parts can exist independently. So each object of an aggregate can be projected into a set of objects of constituent parts. Formally let A be an aggregation of constituent parts B_1, B_2, ..., and B_n. For $e \in A$, the projection of e to B_i is denoted by $e\downarrow_{Bi}$. Then we have

$$(e\downarrow_{B1}) \in B_1, (e\downarrow_{B2}) \in B_2, ..., (e\downarrow_{Bn}) \in B_n.$$

A class aggregated from fuzzy constituent parts must be fuzzy. If the former is still called aggregate, the aggregation is fuzzy. At this point, a class is an aggregation of constituent parts with membership degree of [0, 1]. Correspondingly, we have the following method for determining fuzzy aggregation relationship:

- for any one (fuzzy) object, if the membership degree that it belongs to the aggregate is less than or equal to the membership degree to which its projection to each constituent part belongs to the corresponding constituent part, and
- the membership degree that it belongs to the aggregate is greater than or equal to the given threshold.

The aggregate is then an aggregation of the constituent parts with the membership degree, which is the minimum in the membership degrees to which the projections of these objects to these constituent parts belong to the corresponding constituent parts.

Definition. Let A be an fuzzy aggregation of fuzzy class sets B_1, B_2, …, and B_n, which object membership degrees are μ_A, μ_{B1}, $\mu_{B2, …}$, and μ_{Bn}, respectively. Let β be a given threshold. Then

$$(\forall\ e)\ (e \in A \wedge \beta \le \mu_A\ (e) \le \min\ (\mu_{B1}\ (e\!\downarrow_{B1}), \mu_{B2}\ (e\!\downarrow_{B2}), …, \mu_{Bn}\ (e\!\downarrow_{Bn}))).$$

That means a fuzzy class A is the aggregate of a group fuzzy classes B_1, B_2, …, and B_n if for any (fuzzy) instance object, if the membership degree that it belongs to class A is less than or equal to the member degree to which its projection to anyone of B_1, B_2, …, and B_n, say B_i ($1 \le i \le n$), belongs to class B_i. Meanwhile for any one (fuzzy) object, the membership degree that it belongs to class A is greater than or equal to the given threshold.

Now let us consider the first level of fuzziness in the above-mentioned classes A, B_1, B_2, …, and B_n, namely, they are the fuzzy classes with membership degrees.

Definition. Let A WITH *degree_A* DEGREE, B_1 WITH *degree_B$_1$* DEGREE, B_2 WITH *degree_B$_2$* DEGREE, ……, B_n WITH *degree_B$_n$* DEGREE be classes. Then A is an aggregate of B_1, B_2, …, and B_n if

$$(\forall\ e)\ (e \in A \wedge \beta \le \mu_A\ (e) \le \min\ (\mu_{B1}\ (e\!\downarrow_{B1}), \mu_{B2}\ (e\!\downarrow_{B2}), …, \mu_{Bn}\ (e\!\downarrow_{Bn})) \wedge$$
$$degree\_A \le \min\ (degree\_B_1, degree\_B_2, …, degree\_B_n)).$$

Here β is a given threshold.

It should be noticed that the assessment of fuzzy aggregation relationships given above is based on the extensional viewpoint of class. Clearly these methods can not be used if the classes are defined with the intensional viewpoint because there is no any object available. In the following, we present how to determine fuzzy aggregation relationship using the inclusion degree.

Definition. Let A be an fuzzy aggregation of fuzzy class sets B_1, B_2, …, and B_n and β be a given threshold. Also let the projection of A to B_i is denoted by $A\!\downarrow_{Bi}$. Then

$$\min\ (\mu\ (B_1, A\!\downarrow_{B1}), \mu\ (B_2, A\!\downarrow_{B2}), …, \mu\ (B_n, A\!\downarrow_{Bn})) \ge \beta.$$

Being the same as the fuzzy generation, here $\mu\ (B_i, A\!\downarrow_{Bi})$ ($1 \le i \le n$) means the degree to which B_i semantically includes $A\!\downarrow_{Bi}$. The membership degree that A is an aggregation of B_1, B_2, …, and B_n is $\min\ (\mu\ (B_1, A\!\downarrow_{B1}), \mu\ (B_2, A\!\downarrow_{B2}), …, \mu\ (B_n, A\!\downarrow_{Bn}))$.

Furthermore, the expression above can be extended for the situation that A, B_1, B_2, …, and B_n may have the first level of fuzziness, namely, they may be the fuzzy classes with membership degrees.

Definition. Let β be a given threshold and *A* WITH *degree_A* DEGREE, *B₁* WITH *degree_B₁* DEGREE, *B₂* WITH *degree_B₂* DEGREE, ... *Bₙ* WITH *degree_Bₙ* DEGREE be classes. Then *A* is an aggregate of B_1, B_2, ..., and B_n if

$$\min (\mu\, (B_1, A\!\downarrow_{B1}),\ \mu\, (B_2, A\!\downarrow_{B2}),\ ...,\ \mu\, (B_n, A\!\downarrow_{Bn})) \geq \beta \wedge degree\_A \leq \min$$
$$(degree\_B_1, degree\_B_2, ..., degree\_B_n)).$$

A dashed open diamond is used to denote a fuzzy aggregate relationship. A fuzzy aggregation relationship is shown in Figure 4.8. There a car is aggregated by engine, interior, and chassis. The engine is old and we hereby have a fuzzy class *Old Engine* with the second level of fuzziness. Class *Old Car* aggregated by classes *interior* and *chassis* and fuzzy class *old engine* is a fuzzy one with the second level of fuzziness.

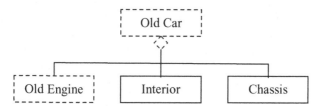

Fig. 4.8. A fuzzy aggregation relationship in the fuzzy UML

4.3.4 Fuzzy Association

Two levels of fuzziness can be identified in the association relationship. The first level of fuzziness means that an association relationship fuzzily exists in two associated classes, namely, this association relationship occurs with a degree of possibility. Also it is possible that it is unknown for certain if two class instances respectively belonging to the associated classes have the given association relationship although this association relationship must occur in these two classes. This is the second level of fuzziness in the association relationship and is caused by such facts that

- the class instances belong to the given classes with possibility degree, and
- the role name of the association relationship is fuzzily defined and there is a fuzzy association with membership degree between two connected instances (e.g., the degree of "friendness" of two people).

Also it is possible that two levels of fuzziness mentioned above may occur in association relationship simultaneously. That means that two classes have fuzzy association relationship at class level on one hand. On the other hand, their class instances may have fuzzy association relationship at class instance level. Note that there may be one kind of fuzzy association modeled by linguistic labels. According to its semantics, the linguistic association should fall into one of the two levels of fuzziness in the fuzzy association relationship given above. So the linguistic association is not discussed.

We can place a pair of words WITH *mem* DEGREE ($0 \leq mem \leq 1$) after the role name of an association relationship to represent the first level of fuzziness in the association relationship. We use a double line with an arrowhead to denote the second level of fuzziness in the association relationship.

Figure 4.9 shows two levels of fuzziness in fuzzy association relationships. In (a), it is uncertain if CD player is installed in car and the possibility is 0.8. So classes *CD Player* and *Car* have association relationship *installing* with 0.8 possibility degree. In (b), it is certain that CD player is installed in car and the possibility is 1.0. Classes *CD Player* and *Car* have association relationship *installing* with 1.0 possibility degree. But at the level of instances, there exits the possibility that the instances of classes *CD Player* and *Car* may or may not have association relationship *installing*. In (c), two kinds of fuzzy association relationship in (a) and (b) arise simultaneously.

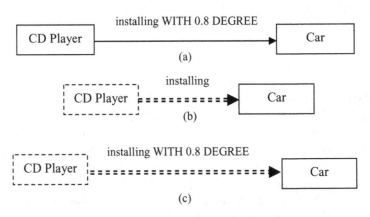

Fig. 4.9. Fuzzy association relationships in the fuzzy UML

The classes with the second level of fuzziness generally result in the second level of fuzziness in the association if this association definitely exists (that means there is no first level of fuzziness in the association). Formally, let A and B be two classes with the second level of fuzziness. Then the instance e of A is one with membership degrees $\mu_A(e)$ and the instance f of B is one with membership degrees $\mu_B(f)$. Assume the association relationship between A and B, denoted $ass\ (A, B)$, is one without the first level of fuzziness. It is clear that the association relationship between e and f, denoted $ass\ (e, f)$, is one with the second level of fuzziness, i.e., with membership degree, which can be calculated by

$$\mu\ (ass\ (e, f)) = \min\ (\mu_A\ (e), \mu_B\ (f)).$$

The first level of fuzziness in the association relationship can be indicated explicitly by the designers even if the corresponding classes are crisp. Assume that A and B are two crisp classes and $ass\ (A, B)$ is the association relationship with the first level of fuzziness, denoted $ass\ (A, B)$ WITH $degree\_ass$ DEGREE. At this moment, $\mu_A\ (e) = 1.0$ and $\mu_B\ (f) = 1.0$. Then

$$\mu\ (ass\ (e, f)) = degree\_ass$$

he classes with the first level of fuzziness generally result in the first level of fuzziness of the association if this association is not indicated explicitly. Formally, let A and B be two classes only with the first level of fuzziness, denoted A WITH $degree\_A$ DEGREE and B WITH $degree\_B$ DEGREE, respectively. Then the association relationship between A and B, denoted $ass\ (A, B)$, is one with the first level of fuzziness, namely, ass (A, B) WITH $degree\_ass$ DEGREE. Here $degree\_ass$ is calculated by

$$degree\_ass = \min\ (degree\_A, degree\_B).$$

For the instance e of A and the instance f of B, in which $\mu_A\ (e) = 1.0$ and $\mu_B\ (f) = 1.0$, we have

$$\mu\ (ass\ (e, f)) = degree\_ass = \min\ (degree\_A, degree\_B).$$

Finally, let us focus on the situation that the classes are ones with the first level and the second level of fuzziness and there is an association relationship with the first level of fuzziness between these two classes, which is explicitly indicated. Let A and B be two classes with the first level of fuzziness, denoted A WITH $degree\_A$ DEGREE and B WITH $degree\_B$ DEGREE, respectively. Let $ass\ (A, B)$ be the association relationship with the first level of fuzziness between A and B, which is explicitly indicated with WITH $degree\_ass$ DEGREE. Also let the instance e of A be with membership degrees $\mu_A\ (e)$, and the instance f of B be with membership degrees $\mu_B\ (f)$. Then we have

$$\mu\ (ass\ (e, f)) = \min\ (\mu_A\ (e), \mu_B\ (f), degree\_A, degree\_B, degree\_a).$$

4.3.5 Fuzzy Dependency

The dependency between the source class and the target class is only related to the classes themselves and does not require a set of instances for its meaning. Therefore, the second level fuzziness and the third level of fuzziness in class do not affect the dependency relationship. A fuzzy dependency relationship is a dependency relationship with degree of possibility, which can be indicated explicitly by the designers or be implied implicitly by the source class based on the fact that the target class is decided by the source class. Assume that the source class is a fuzzy one with the first level of fuzziness. Then the target class must be a fuzzy one with the first level of fuzziness. The degree of possibility that the target class is decided by the source class is the same as the possibility degree of source class.

Let *Employee* and *Employee Dependent* be two classes. It is clear that *Employee Dependent* is dependent on *Employee* and there is a dependency relationship between them. But it is possible that *Employee* may have the first level of fuzziness, for example, with 0.85 possibility degree. Correspondingly, *Employee Dependent* also has the first level of fuzziness with 0.85 possibility degree. Then the dependency relationship between *Employee* and *Employee Dependent* is a fuzzy one with 0.85 degree of possibility.

Since the fuzziness of dependency relationship is denoted implicitly by first level of fuzziness of the source class, a dashed line with an arrowhead can still be used to denote the fuzziness in the dependency relationship. Figure 4.10 shows a fuzzy dependency relationship.

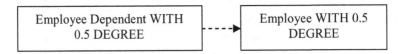

Fig. 4.10. A fuzzy dependency relationship in the fuzzy UML

4.4 Summary

This chapter has presented the fuzzy extensions to ER/EER and UML data models to cope with fuzzy as well as complex objects in the real world at a conceptual level. Some major notions in these data model have been extended and the corresponding graphical representations were developed. It is not difficult to see that a classical conceptual data model is essentially a subset of the corresponding fuzzy conceptual data model. When there is not any fuzziness in the universe of discourse, the latter can be reduced to the former.

The fuzzy ER/EER and UML data models can be applied for conceptual modeling of imprecise and uncertain engineering information in addition to the conceptual design of the fuzzy databases. But limited by their power in engineering information modeling, some data models for engineering should be extended fuzzily for imprecise and uncertain engineering information modeling.

References

Booch, G., Rumbaugh, J. and Jacobson, I. (1998), The unified modeling language user guide, Addison-Welsley Longman, Inc.

Chaudhry, N., Moyne, J. and Rundensteiner, E. A. (1999), An extended database design methodology for uncertain data management, Information Sciences, 121: 83-112.

Chen, G. Q. and Kerre, E. E. (1998), Extending ER/EER concepts towards fuzzy conceptual data modeling, Proceedings of the 1998 IEEE International Conference on Fuzzy Systems, 2: 1320-1325.

Chen, P. P. (1976), The entity-relationship model: toward a unified view of data, ACM Transactions on Database Systems, 1 (1): 9-36.

Galindo, J., Urrutia, A., Carrasco, R. A. and Piattini, M. (2004), Relaxing constraints in enhanced entity-relationship models using fuzzy quantifiers, IEEE Transactions on Fuzzy Systems, 12 (6): 780-796.

OMG (2003), Unified modeling language (UML), http://www.omg.org/technology/documents/formal/uml.htm.

Parsons, S. (1996), Current approaches to handling imperfect information in data and knowledge bases, IEEE Transactions on Knowledge Data Engineering, 8: 353–372.

Ruspini, E. (1986), Imprecision and uncertainty in the entity-relationship model, Fuzzy Logic in Knowledge Engineering, Verlag TUV Rheinland.

Vandenberghe, R. M. (1991), An extended entity-relationship model for fuzzy databases based on fuzzy truth values, Proceedings of the 1991 International Fuzzy Systems Association World Congress, 280-283.

Vert, G., Morris, A., Stock, M. and Jankowski, P. (2000), Extending entity-relationship modeling notation to manage fuzzy datasets, Proceedings of 8th International Conference on Information Processing and Management of Uncertainty in Knowledge-Based Systems, 1131-1138.

Zvieli, A. and Chen, P. P. (1986), Entity-relationship modeling and fuzzy databases, Proceedings of the 1986 IEEE International Conference on Data Engineering, 320-327.

5 The Fuzzy IDEF1X Models

5.1 Introduction

Nowadays manufacturing industries are typically information-based enterprises and computer technology has been involved in nearly all aspects of manufacturing. Computer integrated manufacturing (CIM) is a natural outgrowth of the evolution of computer inclusion in manufacturing activities (Rembold *et al.*, 1993). Led by the United State Air Force, Integrated Computer-Aided Manufacturing (ICAM) is a significant project in computerized manufacturing in the Unite States. ICAM has resulted in several methodologies such as IDEF0, IDEF1/IDEF1X and IDEF3 (Ang and Gay, 1993; Chadha *et al.*, 1991; Doumeingts *et al.*, 1992; IDEF, 2000; Kusiak and Zakarian, 1996), which are respectively applied for building functional models, data models, and process models.

IDEF1X (Appleton, 1986; Kusiak *et al.*, 1997; Loomis, 1986; Wizdom Systems Inc., 1985) has provided a formal framework for consistent modeling of the data necessary for the integration of various functional areas in CIM. The basic idea has been extensively applied in current manufacturing industry. Following on the footsteps of (Zvieli and Chen, 1986), we extend the IDEF1X concepts to model fuzzy data in this chapter. Since the constructs of IDEF1X contain entities, attributes, connection relationships, categorizations, and non-specific relationships, the extension to these constructs must be conducted based on fuzzy sets.

5.2 Fuzzy Entities and Fuzzy Entity Instances

When the set of entities in an IDEF1X model is fuzzy, we get fuzzy entity sets. In this case, membership of an entity in the model is a possibility that is accompanied by some uncertainty and the degree of membership is less then one.

Z. Ma: *Fuzzy Database Modeling of Imprecise and Uncertain Engineering Information,*
StudFuzz **195**, 79–88 (2006)
www.springerlink.com

In order to model the entities with degree of memberships in the IDEF1X model, the degree of membership in the graphic representation of an entity construct should be indicated. The *extended entity name* is employed, which is placed above the box representing the entity together with the entity number and the notation '/'. The syntax of the extended entity name is shown as follows.

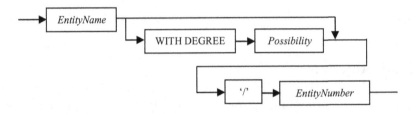

If the degree of membership is an exact one, the parts of WITH DEGREE and the degree can be omitted. At this moment, the fuzzy entities can be reduced to the traditional one. Consider a fuzzy entity CD Player with 0.9 degree of membership in the IDEF1X model. The entity is represented in Figure 5.1.

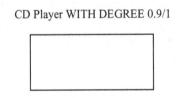

Fig. 5.1. Fuzzy entity with membership degree

There exists another kind of fuzziness of entities in the IDEF1X model, i.e., entity instances fuzzily occur. In other words, entity instances have degree of membership in the corresponding entity. For an entity Academic staff, for example, it is not uncertain if John is instructor.

We use the syntax shown in Figure 5.2 to model the entities that may have fuzzy instances in the IDEF1X model. It should be pointed out that for an entity with fuzzy instances, an additional attribute as a non-key attribute, called membership attribute and denoted by pD, is necessary for the entity to indicate the possibility that each instance belongs to the entity.

It should be noted that two kinds of fuzziness of entities could emerge in either identifier independent entities or identifier dependent entities. But the fuzzy identifier dependent entities are decided by the corresponding fuzzy parent entities.

Academic staff/1

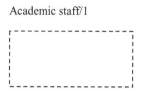

Fig. 5.2. Entity with fuzzy instances

5.3 Fuzzy Attributes and Fuzzy Attribute Values

Being similar to the entities in the IDEF1X model, the fuzziness of attributes can be classified into two categories, namely, fuzzy attributes and fuzzy attribute values. When the set of attributes of an entity is fuzzy, we get fuzzy attribute sets. The membership of an attribute in the entity is a possibility that is accompanied by some uncertainty and the degree of membership is less than one. The fuzzy attribute values mean that the attribute values of the entity instances can take fuzzy values instead of crisp values.

Since the primary/alternative keys or the foreign keys of entities are used to uniquely identify each instance in the entities or the other entities, fuzzy primary/alternative keys or foreign keys as well as fuzzy primary/alternative key values or foreign key values are not permitted. The fuzziness of entities can only emerge in non-key attributes.

After fuzzy attributes and fuzzy attribute values are introduced into IDEF1X model, several constraints on the entity attributes in a fuzzy IDEF1X model should be revised as follows.

- The fuzzy-value rule instead of the no-null rule. Each instance of the entity must have a crisp value for each primary key attribute or foreign key attribute of the entity and may have a fuzzy value for each non-key attribute or alternative key attribute of the entity.
- The fuzzy-functional-dependency rule instead of the full-functional-dependency rule and the no-transitive-dependency rule. Each non-key attribute is functionally dependent on the primary fuzzily. If the primary key is composed of more than one attribute, the value of every non-key attribute is functionally dependent on the entire primary key fuzzily (Ma *et al.*, 2002).

We use the syntax shown as follows to model the fuzzy attributes with degree of memberships in the IDEF1X model.

We use the syntax shown as follows to model the attributes taking fuzzy values in the IDEF1X model.

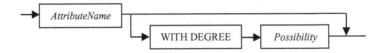

It is possible that an attribute has two kinds of fuzziness simultaneously. For such attributes, the syntax shown as follows is used to represent them.

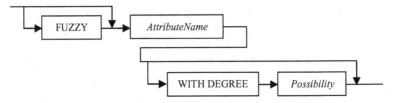

If the degree of membership is an exact one, the parts of WITH DEGREE and the degree can be omitted. At this moment, the fuzzy entities can be reduced to the traditional one. Consider an entity Staff that has an attribute *Weight* with 0.9 degree of membership in the IDEF1X model, and an entity Old-Staff that has an attribute *Age* having fuzzy values. These two entities are represented in Figure 5.3.

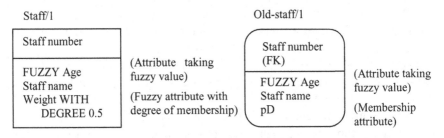

Fig. 5.3. Fuzzy attributes and attributes with fuzzy values

5.4 Fuzzy Connection Relationships

Two levels of fuzziness can be identified in the connection relationships of IDEF1X. When the set of connection relationships in an IDEF1X model is fuzzy, one gets fuzzy connection relationship sets. In this case, membership of a connection relationship between two entities is a possibility that is accompanied by some uncertainty and the degree of membership is less than one. This is the first level. Considering entities Car and CD Player, for example, it has not been decided yet whether to include one connection relationship: (car) install (CD player).

It is possible that although we have decided to include the connection relationship: (car) install (CD player), we do not know whether such connection relationship exists in two entity instances respectively belonging to parent entity and child entity. This case is generally because that the instance is fuzzy with respect to the entity. This is the second level of fuzziness in the connection relationship, which emerges between entity instances.

As we know, there are the identifying relationship and the non-identifying relationship in classical IDEF1X model. Two levels of fuzziness of connection relationship between two entities mentioned above can occur in a non-identifying relationship. Here, either parent entity or child entity can be crisp or fuzzy (fuzzy entity and fuzzy entity instance). But note the situation that the fuzzy entity exists in the first level of fuzziness. Let E_1 and E_2 be fuzzy entities with degrees of membership μ_1 and μ_2, respectively. There is a non-identifying relationship R with the first level of fuzziness, which degree of membership is μ_R. Then we have

$$\mu_R \leq \min (\mu_1, \mu_2).$$

Now let us focus on the identifying relationship. The second level of fuzziness of the connection relationship can occur in an identifying relationship when either parent entity or child entity is crisp or fuzzy. As pointed in Section 3.1, the fuzzy identifier dependent entities are decided by the corresponding fuzzy parent entities. In other words, if the parent entity is a fuzzy entity, then its child entities as identifier dependent entities are fuzzy with the same degrees of membership as that of the fuzzy parent entity. The first level of fuzziness of the identifying relationship only occurs in such situation. At this moment, the degrees of membership of the parent and child entity as well as the connection relationship are identical. That implies that if the parent entity is not a fuzzy one with a degree of membership, the first level fuzziness does not exist in an identifying relationship connected with the entity.

Fuzzy connection relationship is also represented by a line from the parent entity to a child entity. If the connection relationship is with the second level of fuzziness, the line has a small circle at the lower end connecting to the child. If the connection relationship is with the first level of fuzziness, an extended relationship name is placed beside the line. The following syntax is used to represent the extended relationship name.

One could consider the fuzzy connection relationships between entity Car and entity CD Player as well as entity Car engine shown in Figure 5.4.

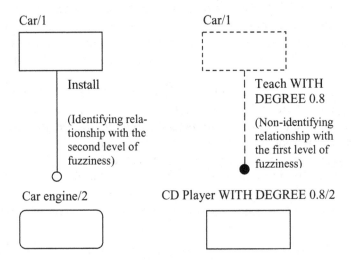

Fig. 5.4. Fuzzy connection relationships

5.5 Fuzzy Non-specific Relationships

Being the same as fuzzy connection relationship, the fuzziness in non-specific relationship can also be identified at two levels. The first level of fuzziness of the non-specific relationship is related to the fuzzy set of the non-specific relationships in the IDEF1X model. Such non-specific relationship is indicated by a relationship name with degree of membership. It is possible that at this moment, two relationship names may have different degrees of membership. The second level of fuzziness of the non-specific relationship occurs because we do not definitely know if two entity instances have any relationship.

A fuzzy non-specific relationship is represented with a line. If the non-specific relationship is with the second level of fuzziness, the line is with a small circle on each end drawn between two entities. If the first level of fuzziness exists in the non-specific relationship, the connection relationship is with the first level of fuzziness, two extended relationship names, which are the same as that in the fuzzy connection relationship, are placed beside the line. They are separated by the notation '/'.

Consider fuzzy non-specific relationship between entity Assembly and entity Part and fuzzy non-specific relationship between entity Staff and entity Project shown in Figure 5.5.

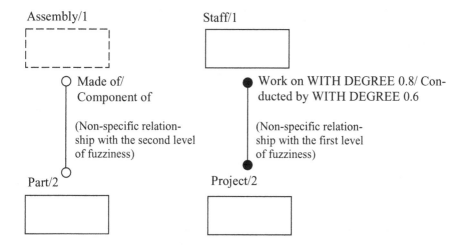

Fig. 5.5. Fuzzy non-specific relationships

5.6 Fuzzy Categorization Relationships

Just like fuzzy identifying relationship, two levels of fuzziness can also been identified in the categorization relationship in an IDEF1X model. The first level of fuzziness of the categorization relationship, being related to the fuzzy set of the categorization relationship in the IDEF1X model, only occurs in the situation that the generic entity is a fuzzy entity, namely, an entity with degree of membership. The degrees of membership of the generic and category entities as well as the categorization relationship are all identical.

The second level of fuzziness of the categorization relationship occurs in the generic entity and its category entities, which are not the fuzzy entities with degrees of membership but can have fuzzy entity instances. At this moment, the fuzzy categorization relationship in involved in the fuzzy instances of the related entities. We mainly focus on this type of fuzzy categorization relationship.

In the classical categorization relationship of IDEF1X, the category entities for a generic entity must be mutually exclusive. For fuzzy categorization relationship, however, the category entities may have fuzzy entity instances. Then one cannot definitely know one instance of the generic entity belongs to which one of the category entities. In other words, since the category entities have fuzzy boundaries, it is possible that the fuzzy category entities for a generic entity may be mutually inclusive. On the other hand, even if the category entities are crisp and they are mutually exclusive, the generic entity may have fuzzy entity instances. One cannot also definitely know such one instance of the generic entity belongs to which one of the category entities.

Based on the discussion above, the complete categorization relationship and the incomplete categorization relationship should be fuzzy and are depicted as follows.

- Fuzzy complete categorization relationship implies that each instance of the generic entity must be contained in one instance of the category entities, but which one instance of the category entities contains it is fuzzy at this moment.
- Fuzzy incomplete categorization relationship implies that a generic entity instance may not be contained in any instance of the category entities.

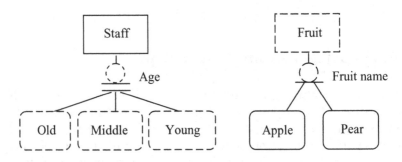

Fig. 5.6. Fuzzy complete versus fuzzy incomplete categorization relationships

A fuzzy categorization relationship is represented with a line from the generic entity to an underlined dashed circle and lines from the dashed circle to each of the category entities. The name of the discriminator is still written next to the dashed circle. For a fuzzy complete categorization, the dashed circle is double underlined. For a fuzzy incomplete categorization, the dashed circle is single underlined.

Fuzzy complete categorization relationship of entity Staff based on attribute Age and fuzzy incomplete categorization relationship of entity Fruit based attribute Fruit name is shown in Figure 5.6.

5.7 Summary

Data modeling is useful in representing and documenting data. It is required that data modeling tool is extended in order to describe the imprecision and uncertainty in engineering design and manufacturing. The IDEF1X technique lends itself to the design and implementation of data models. IDEF1X has gained acceptance in numerous commercial and government establishments.

In this chapter, IDEF1X model has been extended to describe imprecise and uncertain information in the real world. A special focus is on the fuzzy extension. Based on three levels of fuzziness, the basic elements in these three kinds of data models are hereby redefined and the corresponding graphical representations are developed.

References

Ang, C. L., and Gay, R. K. L. (1993), IDEF0 modeling for project risk assessment, Computers in Industry, 22 (1): 31–46.

Appleton (1986), Information Modeling Manual–IDEF1Extended (IDEF1X), D. Appleton Company, Manhattan Beach, CA.

Chadha, B., Jazbutis, G., Wag, C. Y. and Fulton, R. E. (1991), An appraisal of modeling tools and methodologies for integrated manufacturing information systems, Engineering Databases: An Enterprise Resource, 55-63.

Doumeingts, G., Chen, D. and Marcotte, F. (1992), Concepts, models and methods for the design of production management systems, Computers in Industry, 19 (1): 89-111.

IDEF (2000), IDEF Family of Methods, http://www.idef.com/default.html

Kusiak, A. and Zakarian, A. (1996), Reliability evaluation of process models, IEEE Transactions on Components, Packaging, and Manufacturing Technology - Part A, 19 (2): 268 - 275.

Kusiak, A., Letsche, T. and Zakarian, A. (1997), Data modeling with IDEF1X, International Journal of Computer Integrated Manufacturing, 10: 470-486.

Loomis, M. E. (1986), Data Modeling – the IDEF1X Technique, 1986 IEEE Conference on Computers and Communications, 146-151.

Ma, Z. M., Zhang, W. J., Ma, W. Y. and Mili, F. (2002), Data dependencies in extended possibility-based fuzzy relational databases, International Journal of Intelligent Systems, 17 (3): 321-332.

Rembold, U., Nnaji, B. O. and Storr, A. (1993), Computer Integrated Manufacturing and Engineering (Addison Wesley, Reading, MA).

Wizdom Systems Inc. (1985), U.S. Air Force ICAM Manual: IDEF1X, Naperville, IL.

Zadeh, L. A. (1965), Fuzzy sets, Information and Control, 8 (3): 338-353.

Zadeh, L. A. (1978), Fuzzy sets as a basis for a theory of possibility, Fuzzy Sets and Systems, 1 (1): 3-28.

Zvieli, A. and Chen, P. P. (1986), Entity-relationship modeling and fuzzy databases, Proceedings of the 1986 IEEE International Conference on Data Engineering, 320-327.

6 The Fuzzy EXPRESS Model

6.1 Introduction

EXPRESS has been developed as part of the STEP standard for product data modeling and exchange and has been used in a variety of other large scale modeling applications (ISO IS 10303-1 TC184/SC4, 1994; Schenck and Wilson, 1994). The purpose of the information modeling with EXPRESS, in general, centers on the descriptions of the objects that you create to represent information of interest to you. The main objects to be described include the *data types* and *declarations* as well as the *expressions*, *executable statements*, and *interfacing*. In addition, EXPRESS provides a graphical representation of EXPRESS, called *EXPRESS-G*, which uses graphical symbols to conceptually design an EXPRESS model and form a diagram. Note that EXPRESS-G can only represent a subset of the full language of EXPRESS.

Although EXPRESS provides a rich set of constructs to enable complex object definitions, it lacks the ability to model imprecise and uncertain information. With the current edition of EXPRESS, null values are permitted in array data types and role names by utilizing the keyword *Optional*. The application of three-valued logic (False, Unknown, and True), on the other hand, is just a result of the null value occurrences. In addition, the select data types define named collections of other types. An attribute or variable could therefore be one of several possible types. In this context, a select data type also defines one kind of imprecise and uncertain data type whose actual type is unknown to us at present. However, further investigations on the issues of the semantics, representation and manipulation of imprecise and uncertain information in EXPRESS are needed.

Fuzzy information can be found in many engineering activities. To the purpose of fuzzy information modeling with EXPRESS, the constructs of EXPRESS will be extended based on fuzzy sets and possibility distributions in this chapter, including basic elements, various data types, EXPRESS declarations, calculation and operations, and EXPRESS-G.

Z. Ma: *Fuzzy Database Modeling of Imprecise and Uncertain Engineering Information*,
StudFuzz **195**, 89–136 (2006)
www.springerlink.com

6.2 Fuzziness in Basic Elements

The basic elements of EXPRESS consist of *character set, remarks, symbols, reserved words, identifiers* and *literals*. Imprecise and uncertain EXPRESS, being the extension of EXPRESS, should extend these elements to meet the requirement for modeling imprecise and uncertain information. It is therefore necessary that the basic elements of extended EXPRESS should still cover those existing powers in EXPRESS.

Among the basic elements of extended EXPRESS, some, including character set, remarks, symbols and identifiers, are the same as EXPRESS. In the following, the focus is mainly on elements which are different from EXPRESS. The corresponding elements of the original EXPRESS can be found in literature (ISO IS 10303-1, 1994; Schenck and Wilson, 1994).

6.2.1 Reserved Words

The reserved words of EXPRESS include *keywords, operators* and *the names of standard constants, functions* and *procedures*. In the imprecise and uncertain extension of EXPRESS, all reserved words in EXPRESS are included. In addition, several new words are introduced, and they are Fuzzy, With, and Degree.

6.2.2 Literals

A literal is a self-defining constant value. The type of a literal depends on how characters are composed to form a token. The literal types are *binary, integer, real, string* and *logical*. In the following, we present various fuzzy literals.

Binary literal. A binary literal is a token that represents a value of a binary data type and is written as a % followed by one or more binary digits (0 or 1). In a fuzzy binary literal, each binary digit is connected with degree of membership, which means that it fuzzily appears in the corresponding position.

Integer literal. An integer literal is a token that represents a value of an integer data type and is written as a string of digits. Being not the same as a fuzzy binary literal whose fuzziness arises in digits, fuzziness of a fuzzy integer literal is found in strings of digits. In other words, a fuzzy integer literal is a fuzzy set or a possibility distribution, whose support is a set of classical integer literals and each element of the support is connected with degree of membership.

Syntax: Binary literal

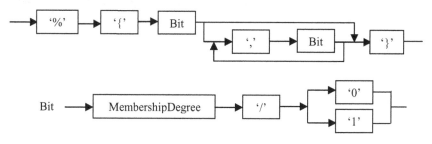

Syntax: Integer literal

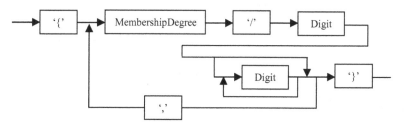

Real literal. A real literal is a token that represents a value of a real data type and is written as a mantissa with a decimal point and an optional exponent. Being the same as a fuzzy integer literal, a fuzzy real literal is a fuzzy set or a possibility distribution, whose support is a set of classical real literals and each element of the support is connected with degree of membership.

Syntax: Real literal

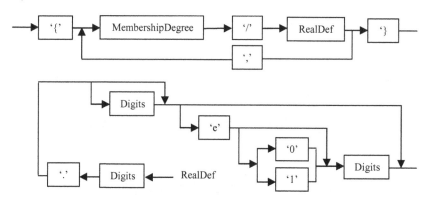

String literal. A string literal is a token that represents a value of a string data type and is written as a string of characters enclosed by quote marks (apostrophes). If a quote mark is in itself part of the string, then two consecutive quote marks are written. Note that a new line character is not legal within a string literal and therefore never spans a physical line boundary. Being the same as a fuzzy binary literal, a string literal may be a fuzzy one, in which each character is connected with degree of membership, which means that it fuzzily appears in the corresponding position.

Syntax: String literal

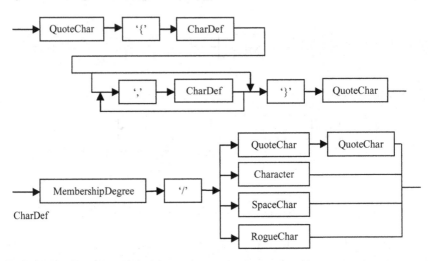

Logical literal. A logical literal is a token that represents a value of a logical or boolean data type, and is written literally as a string of True, False or Unknown. Unknown is not a legal value of a boolean data type. Note that a fuzzy logical literal is a fuzzy set on universe {False, True}. Unknown is not included because it can be represented by a fuzzy set on universe {False, True} which is more informative than Unknown. Unknown can be expressed as {1.0/False, 1.0/True}.

Syntax: Logical literal

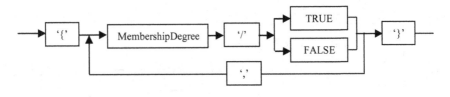

Example 6.1: Here are some fuzzy literals.
%{0.2/1, 0.4/0, 0.7/1, 0.9/1, 0.6/1, 0.3/0}   (*This is a fuzzy binary literal*)
{0.3/15, 0.5/16, 0.8/17, 0.6/18, 0.4/19, 0.1/20} (*This is a fuzzy integer literal*)
{0.5/0.89, 0.7/0.90, 0.9/0.94, 1.0/0.96, 1.0/0.99} (*This is a fuzzy real literal*)
'{0.6/C, 0.7/A, 0.9/C, 1.0/D, 0.8/0, 0.7/0}' (*This is a fuzzy string literal*)
{0.8/True, 0.6/False} (*This is a fuzzy logical literal*)

Aggregate literal. An aggregate literal is a value of an aggregate data type (array, bag, list or set), and is written as zero or more comma separated expressions that evaluate to values compatible with the base type of the aggregate. All of this is enclosed within square brackets. Empty positions are denoted by '?' to represent a 'null'. The numbers of expression values that appear between the brackets have to agree with the bound specification given for the aggregate. The fuzziness of an aggregate literal comes form fuzziness of the base type of the aggregate. In other words, the expressions may evaluate fuzzy values compatible with the fuzzy base type of the aggregate. Being the same as a classical aggregate literal, a repetition factor may be used when several consecutive fuzzy values are the same.

Syntax: Aggregate literal

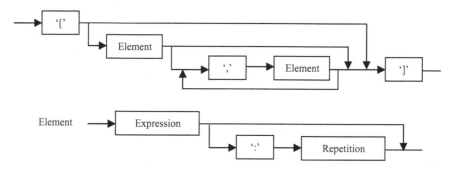

Example 6.2: Here is a set literal which base type is fuzzy integer.
[5, {0.6/7, 0.8/9, 1.0/10}: 2, 12, {0.5/13, 0.9/14, 0.4/15}]

Entity literal. An entity literal is a value of a given entity data type and is written as the entity name followed by comma separated list of expressions enclosed by parentheses. Each value represents an explicit attribute value given in the same order in which they were declared. In a classical entity literal, explicit attributes declared to allow optional values might be given a 'null' value by using '?'.

As it is to be shown in entity data type below, there is fuzziness in entity structures and entity data type itself. This is the first level of fuzziness in a fuzzy entity literal. At this point, an attribute fuzzily belongs to the entity and degree of membership is hereby designated. In addition, an entity data type itself fuzzily belongs to a schema and degree of membership is hereby designated. In addition, an entity literal is fuzzy for the given entity data type; i.e., the entity literal is followed by degree of membership. This is the second level of fuzziness in a fuzzy entity literal. Finally, an explicit attribute value can be fuzzy. This is the third level of fuzziness in a fuzzy entity literal. It can be seen that for an entity literal, only the second fuzziness and the third fuzziness are considered.

Syntax: Entity literal

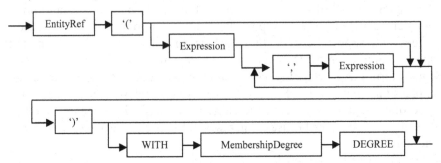

Example 6.3: Assume there is an entity type *poo* with three integer attributes. Two instances are as follows.

poo (5, {0.3/8, 0.6/9, 0.9/10, 0.7/11, 0.5/12}, 15) WITH 1.0 DEGREE

(*Because the membership degree is 1, it can be expressed as *poo* (5, {0.3/8, 0.6/9, 0.9/10, 0.7/11, 0.5/12}, 15)*)

poo (7, {0.6/13, 0.9/14, 0.7/15, {0.8/20, 1.0/21, 0.9/22}}) WITH 0.9 DEGREE

6.3 Fuzzy Data Type Modeling with EXPRESS

Data types are used to represent domains of instance values. A domain is a set of possible values associated with an attribute, local variable, or formal parameter. Data types can be classified as either *Pseudo types* (*Generic, Aggregate*), *Simple types* (*Integer, String,* etc.), *Collection types* (*Array, List,* etc.), *Enumeration types, Select types,* or *Named types* (*Defined types, Entity types*). As shown in Table 6. 1, there are restrictions on the way certain types may be used. It can be seen that pseudo types can only be used as formal parameters and the enumeration and select types can only be used as the underlying types of defined types.

Although EXPRESS has rich data types for data modeling, but a value associated with an attribute, local variable, or formal parameter is only a crisp one in the corresponding domain. In fact, it is possible that the value is imprecise or uncertain on the universe of discourse. With the current edition of EXPRESS, null values are permitted in array data types and role names by utilizing the keyword *Optional*. The application of 3-valued logic (*False, Unknown,* and *True*), on the other hand, is just a result of the null value occurrences. In addition, select data types define named collections of other types. An attribute or variable could therefore be one of several possible types. In this context, a select data type also defines one kind of imprecise and uncertain data type whose actual type is unknown to us at present. It is clear that the set of possible values associated with an attribute, local variable or formal parameter indicates not only data types but also data formats, namely, crisp data or imperfect data. Therefore fuzzy data types should be declared explicitly. Further investigations on the issues of the semantics and representation of fuzzy data types in EXPRESS are needed. Generally speaking, we can put an optional word *FUZZY* before the declaration of data type in the current edition of EXPRESS to indicate if or not the corresponding data type is a fuzzy one.

Table 6.1. Use of data types in EXPRESS

Data type	Attribute	Variable	Parameter	Underlying
Pseudo types			√	
Simple types	√	√	√	√
Collection types	√	√	√	√
Enumera- tion/Select types				√
Named types	√	√	√	√

6.3.1 Pseudo Types

Pseudo types are used only as the types of the formal parameters of functions and procedures. They can be regarded as templates into which various specific types can be placed.

Generic type. The domain of a generic pseudo type is every conceivable value. When a procedure or function with a generic type parameter is invoked, it will accept any kind of actual parameters. This means that fuzzy data is also allowed.

Aggregate type. The domain of an aggregate pseudo type includes any kind of aggregate value. When a procedure or function that has an aggregate type parameter is invoked, it will expect an actual parameter that is an array, bag, list or set. Here, the actual parameter can be fuzzy, depending on whether its base type is fuzzy.

6.3.2 Simple Data Types

Number data type. A number data type is a supertype of the real and integer data types. Its domain is all numbers and is used when it is not known whether the actual representation type is an integer or real. When a value on such domain is not precisely known or obtained, in other words, the domain may be a fuzzy subset or a set of fuzzy subsets of the universe integer or real numbers, *fuzzy number data type* is produced.

Syntax: Number type

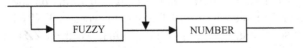

Real data type. A real data type is a subtype of the number data type that represents rational and irrational numbers. Its domain is all real numbers. A real number is represented by a mantissa with an exponent. The number of significant digits in the mantissa is optionally given by PrecisionSpec. Based on the same consideration above, a value of real data type may be fuzzy because its domain may be a fuzzy subset or a set of fuzzy subsets of the universe real numbers.

Syntax: Real type

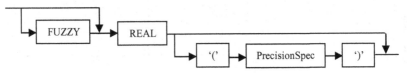

Integer data type. An integer data type is a subtype of the number data type that represents a value of an integer number. Its domain is whole numbers. Note that the number of digits in an integer data type is not constrained in the same manner as a real data type. However, the domain can be constrained either by creating a type with a constraint or by adding a local rule to an entity declaration. A value of integer data type may be fuzzy because its domain may be a fuzzy subset or a set of fuzzy subsets of the universe integer numbers.

Syntax: Integer type

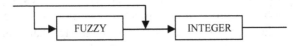

Example 6.4: This example shows some arithmetic operators on fuzzy numeric data.

```
LOCAL
      x : REAL;
      y : FUZZY REAL;
      z : FUZZY REAL (3);
END_LOCAL;
TYPE MyInteger : INTEGER;
                  (* the constraint is put in a type declaration *)
```

```
WHERE
    {0 <= Self <= 50};
END_TYPE;
ENTITY e1;
    a1 : FUZZY MyInteger;
    ...
END_ENTITY;
ENTITY e2;
    a2 : FUZZY INTEGER;

    ...
WHERE                    (* the constraint is put in a local rule *)
    {0 <= supp (a2) <= 50};
END_ENTITY;

    ...
    y := {0.5/0.89225, 0.7/0.90443, 0.9/0.94385, 1.0/0.96476, 1.0/0.99145,
1.0/1.0};
    x := 1/3;
    z := y;   (* z is {0.5/0.89, 0.7/0.90, 0.9/0.94, 1.0/0.96, 1.0/0.99, 1.0/1.0}*)
```

Logical data type. A logical data type represents that the domain values are false, unknown and true, where *False < Unknown < True*. A fuzzy logical value is a fuzzy value on the universe {False, True} with degree of membership. Note that fuzzy logical data types are compatible with fuzzy Boolean data types, which is not the same as that for EXPRESS.

Syntax: Logical type

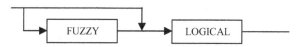

Boolean data type. A boolean data type represents that the domain values are false and true, where *False < True*.

Syntax: Boolean type

String data type. A string data type represents a sequence of zero or more characters. The domain of a string is every permutation of characters, where characters are defined by EXPRESS character set (for all practical purposes, the ASCII character set will do). The case (upper or lower) of letters within a string is significant. A string may be defined as either fixed length (using the Fixed reserved word in the declaration) or varying length. In addition, the maximal number of characters that a string can hold can be limited. In a classical string, it is certain that a character appears in the corresponding position. If it is fuzzy, a fuzzy string is produced. In a fuzzy string, every character has a degree of membership.

Syntax: String type

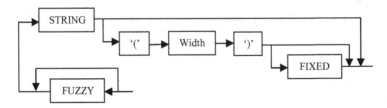

Example 6.5: This example shows some fuzzy string data.
s1 :   FUZZY STRING := '{0.7/C, 0.7/A, 0.7/D, 0.3/D, 0.3/A, 0.3/T, 0.3/A}';
s2 :   FUZZY STRING (10) := '{0.8/E, 0.8/X, 0.8/P, 0.8/R, 0.8/E, 0.8/S, 0.8/S}';
s3 :   FUZZY STRING (9) FIXED := '{0.8/E, 0.8/X, 0.8/P, 0.8/R, 0.8/E, 0.8/S,
0.8/S, 0.3/–, 0.3/G}';

Binary data type. A binary data type represents a sequence of bits (0 or
1's). The domain of a binary is every permutation of bits, where $0 < 1$. Being the same as a string data type, a binary may be defined as either fixed
or variable length and the maximal number of bits that a binary can hold
can be limited. When a bit fuzzily appears in the corresponding position, a
fuzzy binary is produced. In a fuzzy binary, every bit has degree of membership.

Syntax: Binary type

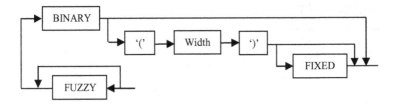

6.3.3 Collection Data Types

Collection types are used to represent ordered or unordered collections.
These collections can have fixed or variable sizes depending on which
specific collection type being considered. Each collection data type has a
different behaviour that suits it to different purposes. The properties of collection data types are shown in Table 6.2. In addition, Collection data
types can be nested to an arbitrary depth, allowing the representation of
any number of dimensions.

The base types of collection data types can be simple data types, named
data types, or another collection data types. Since simple data types and
named data types may be fuzzy, *fuzzy collection data types* are produced
(the fuzziness of named data types will be illustrated latter). In the following declarations of collection data types, the base types can be crisp or

fuzzy simple data types, named data types, or another collection data types.

Table 6.2. Behaviors of collection data types in EXPRESS

	Ordered or Unordered	Fixed or Varying	Duplicate or Un-duplicated
Array data type	Ordered collections	*Fixed sizes*	Duplicate is allowed
Bag data type	Unordered collections	Variable sizes	*Duplicate is allowed*
List data type	Ordered collections	Variable sizes	Duplicate is allowed
Set data type	Unordered collections	Variable sizes	*Duplicate is not allowed*

Array data type. An array data type represents an ordered, fixed-size collection of elements of a given type. The number of elements in an array is fixed by its lower and upper bound, in which both the lower and upper bounds are integer-valued expressions that may be negative, zero or positive and the former is less than or equal to the latter. In EXPRESS, each element of an array data type must be different from any other element in the same array value by using the *Unique* keyword in the declaration. In addition, the elements of an array data type are allowed to have a null value by using the *Optional* keyword in the declaration. The fuzzy elements of an array data type, however, are not considered.

Syntax: Array type

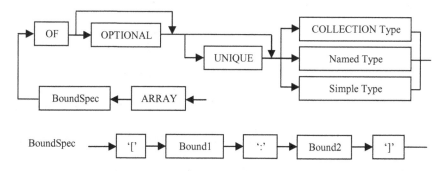

Bag data type. A bag data type represents an unordered collection of like element types within which duplicate element values are permitted. The number of bag elements can be specified if needed. Otherwise, the bag can hold any number of elements. Note that although the position of an

element in a bag is not significant, each of elements will be at certain position and can thereby be accessed by a subscript. After an element is either inserted or deleted, any element in the bag may be in a different position from before. Being not the same as an array data type, however, no element of a bag data type is permitted to be null values in classical EXPRESS. In other words, the *Optional* keyword is not used in the declaration. But it is possible that the elements of an array data type are fuzzy.

Syntax: Bag type

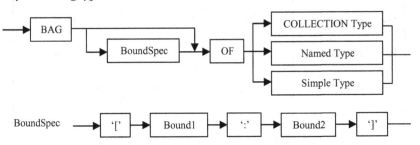

List data type. A list data type represents an ordered collection of like element types. The number of list elements can be specified if needed. Otherwise, the list can hold any number of elements. Duplicate elements are allowed in a list unless it is declared as unique. In the declaration of a list data type, the *Optional* keyword is not used. It is meant that no element of a list data type is permitted to be null values in EXPRESS. But it is possible that the elements of a list data type are fuzzy.

Syntax List type

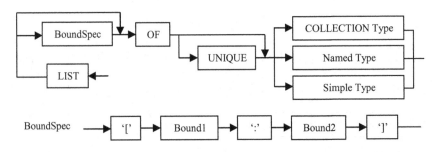

Set data type. The set data type is similar to the bag data type. A set data type represents an unordered collection of like element types. But no two elements of a set can have the same value. The number of set elements can be specified if needed. Otherwise, the bag can hold any number of elements. Although the position of an element in a set is not significant, each of elements will be at certain position and can thereby be accessed by a subscript. After an element is either inserted or deleted, an element in the set may be in a different position from before. In the declaration of a set data type, the *Optional* keyword is not used.

Syntax: Set type

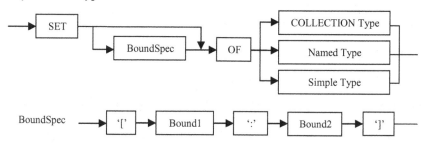

Example 6.6: This example shows some fuzzy collection data types.
ARRAY [0 : 10] OF FUZZY INTEGER;
BAG OF FUZZY REAL;
LIST [0 : ?] OF FUZZY NUMBER;
SET FUZZY STRING (9);

6.3.4 Enumeration Type

An enumeration type is an ordered list of values represented by names. The values of the enumeration type are designated by enumeration items. The order of the values of an enumeration type is determined by their relative position in the enumeration item list: the first occurring item is less than the second; the second is less than the third, etc. Comparison between values in different enumeration types is undefined even if the item names are the same.

Syntax: Enumeration type

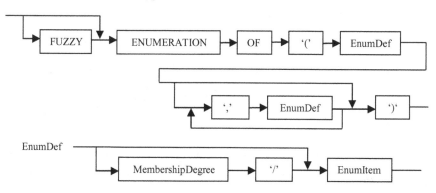

Note that in EXPRESS, an enumeration item belongs only to the type by which the item is defined and must be unique within that type definition.

In other words, the domain defined by an enumeration data type is a crisp set of values represented by names. When this set is a fuzzy one, i.e., each enumeration item of the enumeration data type may be connected with degree of membership, a fuzzy enumeration data type is produced.

6.3.5 Select Type

A select type defines a named collection of other types called a select list. A value of a select data type is a value of one of the types specified in the select list where each item is an entity type or a defined type. This allows an attribute or variable to be one of several possible types. Therefore, the domain of values for such a type is the union of the domains of the types in its select list.

The reason that select data types are introduced is essentially information imprecision and uncertainty. We do not know the perfect data type for an attribute or variable but know a finite set of all possible types at present. In this set, only one value will be true. Such semantics is fully the same as that of partial values. It should be noticed that the possibility that each possible type in the select list is true is equal. In order to represent the different possibility for the different type in the select list, a degree of membership is designated to every type in the select list. A fuzzy select type is hereby produced.

Syntax: Select type

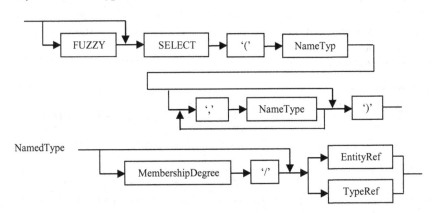

Example 6.7: Here a fuzzy enumeration type and a fuzzy select type are created to represent fuzzy color values and a choice made among fuzzy things, respectively.

TYPE *HairType* = FUZZY ENUMERATION OF (0.1/Black, 0.4/Grey, 0.7/Brown, 0.5/Golden, 0.2/Red);
   END_TYPE;

```
TYPE SnackFood = FUZZY SELECT (0.3/Apple, 0.6/Orange, 0.4/Tomato);
END_TYPE;
...
hair : HairType;
snack : SnackFood;
```

6.4 Fuzzy Declarations

The purpose of EXPRESS, and the information modeling process in general, centers on the declarations of the objects that you create to represent information of interest to you. The main object to be declared is a *schema*. Within a schema, *constants*, *types*, *entities*, *functions*, *procedures*, and *rules* should be declared. Among these things, attributes, local variables and parameters are the sub-objects.

The structure of EXPRESS source defines the scope of declarations. This process of declaring things inside other things indicates the way one declaration 'own' other things. In the case of an entity declaration, the attributes and local rules represent the properties from which an entity declaration is defined. So, an attribute is owned by a particular entity and it is also a property of the entity.

6.4.1 Constant

A constant declaration is used to create values that never change. A constant may appear in the declaration of another constant but circular definitions are not allowed. EXPRESS forces you to declare all constants (in a schema) in the same 'block' before any other declarations are made.

Note that a constant must belong to a base type (simple type, named type, or collection type). It has been shown in the above section, or will be shown in this section, that these base types may be fuzzy. So a constant being to a type may be a fuzzy one.

Example 6.8: Here are a few constants.

```
CONSTANT
data1 : FUZZY REAL := {0.5/1.357, 0.7/1.689, 0.4/1.907};
data2 : FUZZY INGEGER := {0.3/15, 0.8/16, 0.4/17};
data3 : FUZZY STRING := '{0.7/F, 0.7/U, 0.7/Z, 0.7/Z, 0.3/-, 0.3/I, 0.3/E, 0.3/E,
0.3/E}';
END_CONSTANT;
```

6.4.2 Type

A type declaration provides the facility to create a defined type, which is used to distinguish conceptually different collections of values that happen to have similar representations. A defined type is declared based on the underlying type—simple type, collection type, named type, or enumeration type. Generally speaking, the defined type has the same domain of values as the underlying type (unless a constraint is put on it) and the operations that can be performed on it are the same as those applicable to the underlying type.

The fuzziness of a defined type comes from that of its underlying type. In addition, for a defined type, there may be fuzziness in constraints on the domain of the underlying types due to fuzzy Boolean or logical results. The fuzziness in rules will be illustrated later.

Example 6.9: A defined type based on fuzzy real is created. It helps to reveal the meaning.

```
TYPE length = FUZZY REAL;
END_TYPE;
```

6.4.3 Entity

An entity declaration creates a type that defines the properties of real-world or conceptual objects. An entity is either simple or complex. A simple entity is one that is neither a subtype nor a supertype. A complex entity is part of a system (a lattice) of supertypes and subtypes.

The properties of an entity are defined in terms of attributes and constraints. Attributes define the material properties of an entity and always have a value domain. Attributes may have values that are explicit or derived. Constraints are static and may govern any of the properties such as "cardinality constraints," "uniqueness rules," "domain rules," and "global rules."

The scope of an entity declaration is the entity itself plus all of its supertypes (if any). Every identifier created in the entity declaration is unique within the current scope. This includes all attribute names and rule labels appearing within the entity declaration itself and all those inherited from any of its supertypes. It is possible to refine the declaration of an attribute of one of its supertypes by giving it a different type in the subtype declaration.

The fuzziness of an entity can be differentiated at three levels. The first level of fuzziness is on the structure of an entity. In other words, it is not completely confirmed that a property must belong to an entity, i.e., it belongs to the entity with a degree of membership. The second level of fuzziness is on the instances of the entity. Even though the structure of an entity is crisp, it is possible that an instance of the entity belongs to the entity with a degree of membership. On the other hand, an attribute in an entity defines a value domain. When this domain is a fuzzy subset or a set of a fuzzy subset, the fuzziness of an attribute value appears. This is the third level of fuzziness.

Attribute

An attribute of an entity may be explicit, derived, or inverse. The explicit and derived attributes represent its material properties. Inverse attributes represent existential dependency. In all cases, attributes have a domain of values that is defined by its type, including those that are crisp, or those that are imprecise and uncertain. In addition, as mentioned above, an attribute may have a degree of membership.

In order to model the first level of fuzziness in an attribute, i.e., an attribute with a degree of membership, a pair of words WITH *mem* DEGREE is introduced, where $0 \leq mem \leq 1$ and is used to indicate the degree that the attribute belongs to the entity. Generally, an attribute will not be declared when its degree is 0. In addition, "WITH *1.0* DEGREE" can be omitted when the degree of an attribute is 1. One or more attributes with degrees of membership can be declared in the same statement when everything to the right of the colon is identical for all of them, including data type and degree of membership.

If an attribute is allowed to have a null value, the keyword OPTIONAL is used. Similarly, if one wants to allow an attribute to have a fuzzy value (the third level of fuzziness in an attribute), a fuzzy collection type, fuzzy named type, or fuzzy simple type is chosen for it. Note that OPTIONAL and fuzzy data type for an attribute can be employed simultaneously. At this point, the attribute value can be crisp, null, or fuzzy.

Explicit attribute. An explicit attribute is given in the syntax of explicit attribute declaration.

Syntax: Explicit attribute declaration

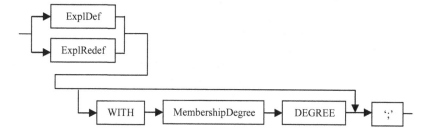

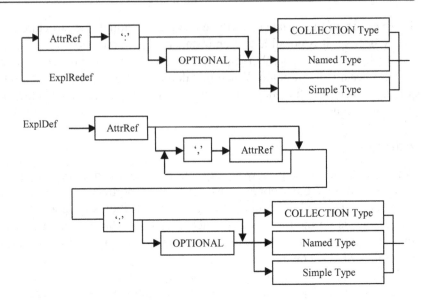

Derived attribute. A derived attribute is a property of an entity whose value changes in response to changes in other attribute values. In EXPRESS, a derived attribute declaration gives a role name, a type, and an expression that defines how its value is derived. Being the same as an explicit attribute, the first and third levels of fuzziness can be found in a derived attribute.

Syntax: Derived attribute declaration

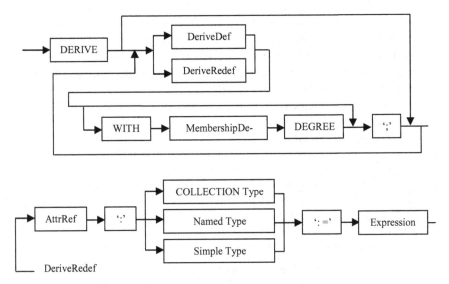

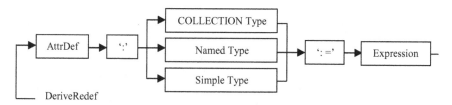

Inverse attribute. An inverse attribute is a property of an entity whose value represents the coupling between a value of the "subject" entity (the one in which the inverse attribute is declared) and values of the "user" entity that refers the subject entity as a type. The value of an inverse attribute is dynamic, as it changes in response to changes that occur elsewhere in the information base. In classical EXPRESS, an inverse attribute declaration gives a role name, the user entity type, and the attribute of the user that refers to the subject entity. The fuzziness of a inverse attribute is decided by that of the attribute of the user that refers to the subject entity.

Example 6.10: A defined type based on fuzzy real is created. It helps to reveal the meaning.

```
ENTITY widget;
      id: STRING;
      length, width, height: FUZZY REAL;
      material: STRING WITH 0.8 DEGREE;
DERIVED
      capacity: FUZZY REAL := length * width * height;
END_ENTITY;
ENTITY customer;
      name: STRING;
      address: STRING;
      phone: OPTIONAL FUZZY STRING;
INVERSE
      ordering: widget FOR id;
END_ENTITY;
```

Local Rule

Local rules are given in the local context of an entity declaration. The local rules are assertions of the validity of entity instances and apply to all instances of that entity type. There are two types of local rules: uniqueness rules that control the uniqueness of attribute values among all instances of a given entity type; and domain rules that describe other constraints on or among the attribute values of each instance of a given entity type. Local rules may optionally be given a label. Labels are used to identify rules in documentation, in error reports, and in enforcement specifications.

In EXPRESS, for a given instance of the declared entity type, a local rule has one of three states: true, false, or unknown. However, a local rule has a fuzzy logical value: fuzzy true or fuzzy false.

Unique rule. Following the UNIQUE keyword, a unique rule specifies either a single attribute name or a list of two or more of them. The former rule is a "simple uniqueness constraint" requiring that an attribute value be unique, and the latter rule is a "joint uniqueness constraint" requiring that a tuple of values be unique. Since an explicit attribute can be optional or fuzzy valued and unique, a uniqueness constraint where an attribute is allowed to have optional and fuzzy values, evaluates to a fuzzy logical value when there is a missing or a fuzzy value.

Domain rule. Domain rules, following the WHERE keyword, are used to specify constraints on the value of individual attributes or a combinations of attributes in an entity value. The domain rule is given by a sequence of logical valued expressions. The ingredients in the expression must come from a single entity value taken in isolation. The expression can reference attributes, constants, and functions as long as those do not refer to other entity declarations. When a domain rule contains an optional or fuzzy-valued attribute, or a fuzzy function (this will be demonstrated in the following), it evaluates to a fuzzy logical value. The syntax for domain rule here is the same as that under a classical condition.

Supertypes and Subtypes

Subtypes and supertypes are used to build a classification structure in which subtypes are more specific than supertypes and supertypes are more general than subtypes. Therefore, every value of a subtype is a value of its supertypes.

However, as previously shown, there are three levels of fuzziness in entities. Despite the first level of fuzziness, i.e., that attributes belong to the corresponding entities with a degree of membership and that the existence of the entities is with a degree of membership, entities may have the attributes with a fuzzy value domain and in that case an entity instance belongs to the entity with a membership degree. Fuzzy entities are thus produced.

Being similar to fuzzy object-oriented databases, fuzzy entities result in a fuzzy supertype–subtype relationship. In other words, an entity becomes the subtype of a fuzzy entity by inheriting all attributes and rules of the latter and adding some new attributes and rules. It is clear that the subtype is fuzzy at this point and the subtype is that of the supertype with a degree of membership. The method introduced in Section 3.4.3 can be used to evaluate such a degree of membership and indicate it in the declaration of a subtype. If a subtype has multiple supertypes, it may have a different degree of membership with respect to each supertype. As to the confusion in multiple inheritances, the same rule as that for resolving conflicts of multiple inheritance in fuzzy object-oriented databases can be used. That is, the subtype inherits the conflicted attributes from the fuzzy supertype that has the highest degree of membership with the subtype.

Note that in order to more precisely measure the semantic relationships between entity instances and the entity as well as subtype and supertype, it is suggested that the weights of attributes should be defined in the declarations of entity just like the definitions of class in object-oriented databases. This demonstrates the need for modeling fuzzy information.

Subtype declaration

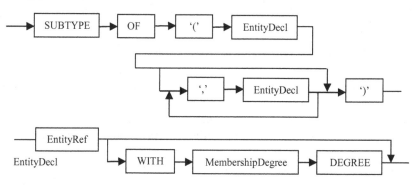

Entity Definition

Based on the discussion above, the syntax of the entity definition is given as follows.

It should be noted that in the definition of entity above, an additional attribute must be added. This attribute, called *membership degree attribute* and written as *pD*, is neither the explicit attribute, nor derived or inverse attribute mentioned above. It is used to represent the possibility that entity instances belong to the entity type.

Syntax: Entity declaration

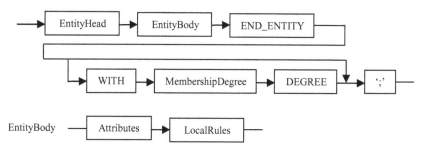

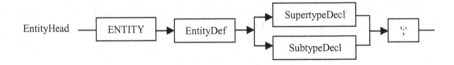

6.4.4 Algorithm

An algorithm is a sequence of statements that perform certain operations. Two kinds of algorithms can be specified in EXPRESS: functions and procedures. Formal parameters define the input to an algorithm. When an algorithm is called, actual parameters supply values. The actual parameters must agree in type, order, and number with the formal parameters. Declarations local to the algorithm are given following the header as needed. These declarations can be types, local variables, other algorithms, etc. The body of the algorithm follows local declarations.

Formal parameter. Formal parameters tell functions and procedures what kind of input is required, and in the case of procedures, what kind of output can be produced. Each parameter has a name and a type. A formal parameter to a procedure may be declared as Var (variable), which means that if its actual value is changed within the procedure, the change is apparent to the point of invocation. Function parameters cannot be variable. Pseudo types (generic or aggregate) or conformant types can be used to generalize the kind of actual values that may be passed to functions and procedures. A conformant aggregate is a form of aggregation that does not specify its bounds. Here the actual parameter can be fuzzy.

Local variable. Local variables of an algorithm are declared inside a local block labeled between LOCAL and END_LOCAL. They are allowed to assign a fuzzy value to a logical variable. For a local variable, the implicit value is a null value unless an explicit initialization is given.

Example 6.11: Let us look at the following example.
```
LOCAL
      i: STRING := 'EXPRESS/STEP';
      j: FUZZY REAL := {0.6/3.6, 0.8/3.8, 1.0/4.0, 0.7/4.2, 0.5/4.4};
      k: INTEGER;
END_LOCAL;
```
Function. A function is an algorithm that operates on its input parameters and returns a single value of the specified type to the point of invocation. Functions may be fuzzy. A fuzzy function means that its input parameters may be fuzzy data types and local variables may also be fuzzy. Accordingly, the returned value of the function may be fuzzy.

Example 6.12: Here is a fuzzy function.
```
FUNCTION root (a, b, c: FUZZY REAL): FUZZY REAL;
LOCAL
      deta: FUZZY REAL;
```

```
END_LOCAL;
    deta := b ** 2 – 4 * a * c;
    IF deta > = 0.0 AND SQRT (deta) – b >= 0 THEN
            root := (SQRT (deta) – b)/(2 * a);
    END-IF;
    RETURN (root);
END_FUNCTION;
```

Procedure. A procedure is an algorithm that receives parameters from the point of invocation and operates on them in some manner to produce the desired goal. Being similar to functions, procedures may also be fuzzy. A fuzzy procedure means that the parameters supplied to the procedure and local variables are fuzzy data types. A procedure is declared inside a block labeled between PROCEDURE and END_PROCEDURE.

6.4.5 Rule

The constraints provided by local rules (i.e., the UNIQUE and WHERE clauses in an entity declaration) apply to every instance. RULE permits the definition of constraints that apply to restricted scopes or that cannot be expressed in a local rule. In a rule, the rule header names the rule and specifies the entities affected by it. A rule is declared inside a block labeled between RULE and END_RULE. Being similar to a local rule, a global rule may have a fuzzy logical value in addition to true or false. Rules may be fuzzy depending on their evaluation.

6.4.6 Schema

A schema, enclosed by SCHEMA and END_SCHEMA, defines a collection of objects that have a related meaning and purpose. Normally a schema declaration contains declarations of constants, entities, functions, procedures, rules, and types, which are those that have been fuzzily extended. The first-level fuzziness of entities, i.e., entities belonging to the schema with a degree of membership, can be embodied in the definition of schema.

6.5 Expressions with Fuzzy Information

Expressions in EXPRESS are combinations of operators, operands, and function calls that are evaluated to produce a crisp value or a fuzzy value. There are two primary kinds of operators. A binary operator requires two operands and is placed between those operands. A unary operator requires only one operand and precedes an operand. Note that the operands are required to be type compatible with the operators.

According to such requirement, the operators in EXPRESS are classified as follows:

- Arithmetic operators accept number operands and produce number results, which type dependent on the operators and the types of the operands.
- Relational operators accept various types of operands and produce Logical (*True*, *False*) results.
- Binary operators accept Binary operands and produce Binary results.
- Logical operators accept Logical operands and produce Logical results.
- String operators accept String operands and produce String results.
- Aggregate operators combine aggregate values with other aggregate values or with individual elements in various ways and produce aggregate results.
- Component reference operators extract components from entity instances and aggregate values.

Table 6.3. Operator precedence in EXPRESS

Precedence	Description	Operators
1	Component references	[], ., \
2	Unary operators	+, −, NOT
3	Exponentiation	**
4	Multiplication/Division	*, /, DIV, MOD, AND
5	Addition/Subtraction	+, −, OR, XOR
6	Relational	=, <>, <=, >=, <, >, :=:, :<>:, IN, LIKE

Evaluation of an expression is governed by the precedence of the operators. If expressions are enclosed by parentheses, they are evaluated before being treated as a single operand. Evaluation proceeds from left to right, with the highest precedence being evaluated first. The precedence rules for all of the operators of EXPRESS are shown in Table 6.3.

6.5.1 Arithmetic Operators

The arithmetic operators can be unary or binary, where the unary arithmetic operators are *identity* (+) and *negation* (−), and the binary arithmetic operators are *addition* (+), *subtraction* (−), *multiplication* (*), *division* (/), *exponentiation* (**), *integer division* (DIV), and *modulo* (MOD).

Following Zadeh's (1975) extension principle, a binary arithmetic operator can produce a fuzzy result when either of the operands is fuzzy. The fuzzy result is a fuzzy integer for operators "+", "−", "*", "/", and "**" if both operands have fuzzy integer data types, or one has a fuzzy integer data type and another has an integer data type, and a fuzzy real result otherwise. For "DIV" and "MOD," the fuzzy result is always a fuzzy integer. If either operand has a (fuzzy) real data type, it is truncated to an (fuzzy) integer before the operation takes place.

A unary operator accepting a fuzzy single numeric operand produces a fuzzy value of the same type. In the case of "+", the result is the same as the operand and the result is the negation of the operand in the case of "−".

Rounding and truncation. For a real number, truncation drops all digits to the right of the decimal point, producing an integer value. Rounding, however, is performed when an (evaluated) expression is assigned to a variable. Given x, which is the precision, the digit $x + 1$ is examined. If that digit is 5 or greater, a carryover of 1 is added to the digit x and all digits beyond x are discarded; otherwise all digits beyond x are simply discarded. For a fuzzy real number, truncation and rounding are performed on all elements in its support, and produce fuzzy integer values.

Example 6.13: This example shows some arithmetic operators on fuzzy numeric data.

```
LOCAL
      a1, a2, b1, b2 : FUZZY INTEGER;
      x1, x2, y1, y2 : FUZZY REAL;
      area : FUZZY REAL (6);
END_LOCAL;
      ...
      a1 := {0.7/1, 0.9/2, 1.0/3};
      x1 := {0.6/4.4, 0.8/5.4, 1.0/6.4};
      b1 := a1 * 3; y1 :=  x1 * 2;
      a2 := x1 DIV 2; b2 := y1 MOD 7;
      x2 := + x1 * 2; y2 := - y1;
      area := PI/4 * ((x1 − x2) ** 2 + (y1 − y2) ** 2);
```

After the operations, b1 and y1 are computed to be {0.7/3, 0.9/6, 1.0/9} and {0.6/8.8, 0.8/10.8, 1.0/12.8}, respectively. Furthermore, a2, b2, x2, and y2 are respectively computed to {0.8/2, 1.0/3}, {0.6/1, 0.8/3, 1.0/5}, {0.6/8.8, 0.8/10.8, 1.0/12.8}, and {0.6/-8.8, 0.8/-10.8, 1.0/-12.8}. In addition, area is computed to be a fuzzy value {0.6/2.58490243... e + 2, 0.6/3.17238026... e + 2, 0.6/3.75985808... e + 2, 0.8/3.89337577... e + 2, 0.8/4.611700093... e + 2, 1.0/5.46888449... e + 2} but has an actual value of {0.6/2.58490 e + 2, 0.6/3.17238 e + 2, 0.6/3.75986 e + 2, 0.8/3.89338 +

2, 0.8/4.61170 e + 2, 1.0/5.46888 e + 2} since the specification calls for six digits of precision.

Note that being the special cases, crisp values and partial values (including intervals) can be treated as fuzzy values in the evaluations of expressions whose degrees of membership are exact. When null values are encountered in expressions, however, null answers are produced.

6.5.2 Relational Operators

The relational operators in EXPRESS expressions can be membership (IN) and string match (LIKE) in addition to value comparison operators and instance comparison operators. The result of a relational expression is a Logical value (True, False, or Unknown). If either operand is a null value, the expression evaluates to Unknown. Generally speaking, when a fuzzy value (including a partial value or a value interval) is either operand, the result of the expression is also Unknown, which is further described by a fuzzy Boolean value, a fuzzy value on the universe of discourse {True, False}. If a threshold value is given, however, the expression can be evaluated to be True or False.

Value Comparison Operators

The value comparison operators are *equal* (=), *not equal* (<>), *greater than* (>), *less than* (<), *greater than or equal* (>=), and *less than or equal* (<=). These operators can be used on numeric, logical, string and enumeration operands. In addition, = and <> can be used for values of aggregate, binary, and entity types.

Numeric Comparisons. The value comparison operators used on fuzzy numeric operands have been investigated in (Ma, Zhang, and Ma, 1999) by utilizing the notion of semantic equivalence degree and the given threshold. Here, the focus is on the fact that the thresholds are not given. At this point, the result of a fuzzy value comparison operation is a fuzzy Logical value. Let A and B be fuzzy values with the membership functions μ_A and μ_B, respectively, and their equivalence degree SE $(A, B) = \gamma$ $(0 \leq \gamma \leq 1)$. Then

- $A = B$: $\{\gamma/\text{True}, (1 - \gamma)/\text{False}\}$
- $A <> B$: $\{(1 - \gamma)/\text{True}, \gamma/\text{False}\}$
- $A > B$: $\{0/\text{True}, 1/\text{False}\}$ when min (supp (A)) $\leq$ min (supp (B)) and max (supp (A)) $\leq$ max (supp (B)), otherwise $\{(1 - \gamma)/\text{True}, \gamma/\text{False}\}$
- $A < B$: $\{0/\text{True}, 1/\text{False}\}$ when min (supp (B)) $\leq$ min (supp (A)) and max (supp (B)) $\leq$ max (supp (A)), otherwise $\{(1 - \gamma)/\text{True}, \gamma/\text{False}\}$

- $A \geq B$: $\{\gamma/\text{True}, (1 - \gamma)/\text{False}\}$ when min (supp (A)) $\leq$ min (supp (B)) and max (supp (A)) $\leq$ max (supp (B)), otherwise $\{\max (\gamma, (1 - \gamma))/\text{True}, (1 - \max (\gamma, (1 - \gamma)))/\text{False}\}$
- $A \leq B$: $\{\gamma/\text{True}, (1 - \gamma)/\text{False}\}$ when min (supp (B)) $\leq$ min (supp (A)) and max (supp (B)) $\leq$ max (supp (A)), otherwise $\{\max (\gamma, (1 - \gamma))/\text{True}, (1 - \max (\gamma, (1 - \gamma)))/\text{False}\}$

Example 6.14: This example shows some value comparison operators used to fuzzy numeric operands.

a : FUZZY INTEGER := {0.5/1, 0.7/2, 0.9/3, 1.0/4, 0.8/5, 0.6/6, 0.4/7};
b : FUZZY INTEGER := {0.5/1, 0.7/2, 0.9/3, 1.0/4, 0.8/5, 0.6/6, 0.4/7};
c : FUZZY INTEGER := {0.6/2, 0.8/3, 1.0/4, 0.9/5, 0.6/6, 0.4/7, 0.1/8};
d : FUZZY INTEGER := {0.3/5, 0.6/6, 0.9/7, 1.0/8, 0.8/9, 0.6/10, 0.4/11};

Utilizing the definition of semantic equivalence degree in Chapter 7, one has

SE (a, b) = min $((0.5 + 0.7 + 0.9 + 1.0 + 0.8 + 0.6 + 0.4)/(0.5 + 0.7 + 0.9 + 1.0 + 0.8 + 0.6 + 0.4), (0.5 + 0.7 + 0.9 + 1.0 + 0.8 + 0.6 + 0.4)/(0.5 + 0.7 + 0.9 + 1.0 + 0.8 + 0.6 + 0.4)) = 1,

SE (b, c) = min $((0.6 + 0.8 + 1.0 + 0.8 + 0.6 + 0.4)/(0.5 + 0.7 + 0.9 + 1.0 + 0.8 + 0.6 + 0.4), (0.6 + 0.8 + 1.0 + 0.8 + 0.6 + 0.4)/(0.6 + 0.8 + 1.0 + 0.9 + 0.6 + 0.4 + 0.1)) = 0.857, and

SE (c, d) = min $((0.3 + 0.6 + 0.4 + 0.1)/(0.6 + 0.8 + 1.0 + 0.9 + 0.6 + 0.4 + 0.1), (0.3 + 0.6 + 0.4 + 0.1)/(0.3 + 0.6 + 0.9 + 1.0 + 0.8 + 0.6 + 0.4)) = 0.304$. Therefore,

a = b is {1/True, 0/False}, i.e., True,
b = c is {0.857/True, 0.143/False}, and
c < d is {(1 − 0.304)/True, 0.304/False}, i.e., {0.696/True, 0.304/False}.

Binary and String Comparisons. String and binary values use the same procedure for comparison, but use different character sets. The character set and collating order for string values is defined by the ordering of the EXPRESS character set; the character set for binary values is 0 and 1 where 0 < 1.

To compare two values, compare the first (leftmost) pair of characters (ignoring their membership degrees), then the pair at the second position, etc., until an unequal pair is found or until all pairs have been examined in the following order:

- If an unequal pair is found, no additional comparison is needed and the value with the lesser character is less than the other. The possibility that the result is true is the minimum of the membership degrees of the two characters in the unequal pair.
- If one value has fewer characters than the other, then it is the lesser. The possibility that the result is true is the maximum of the membership

degrees of the characters that are from the value with more characters and are different from the characters of the other value.

- If both values have the same length and all pairs are equal, the two are equal. The possibility that the result is true is the minimum of the membership degrees of all characters in the two values.

Example 6.15: Let s_1, s_2 and s_3 be fuzzy strings, where s_1 = '{0.4/a, 0.7/b, 0.5/c}', s_2 = '{0.5/a, 0.9/b, 0.6/c}', and s_3 = '{0.3/a, 0.5/b, 0.8/c, 0.5/d, 0.3/e}'. Then the equality s_1 = s_2 is evaluated to be {0.4/True, 0.6/False}, and the inequality $s_2 < S_3$ is evaluated to be {0.5/True, 0.5/False}.

Logical Comparisons. Let two fuzzy logical values {λ_1/True, ζ_1/False} and {λ_2/True, ζ_2/False}. Then {λ_1/True, ζ_1/False} < {λ_2/True, ζ_2/False} if and only if $\lambda_1 < \lambda_2$.

Enumeration Comparisons. Generally speaking, the first occurring item in data of an enumeration type is less than the second, and the second is less than the third, etc. However, each item occurs with a degree of membership, so the relationship of "less than" should be fuzzy. In other words, the results of enumeration comparisons are fuzzy logical values. The possibility of being true for each comparison pair is the minimum of its membership degrees.

Aggregate Comparisons. Aggregate values can be compared for equal (=) and not equal (<>). Two aggregate values can be compared only if their types are compatible, which means that they are both the same kind of aggregate; have the same declared bounds in the case of an array or same actual size in the case of a bag, list or set; and have compatible base types. Considering the elements in aggregate values may be of a fuzzy data type, the definitions of equality are respectively as follows:

- Two arrays x and y are equal if and only if each element of x is equal to the element of y at the same position. Equality $x = y$ is evaluated to be a fuzzy logical value {$\min_i$ (SE (x [i], y [i]))/True, (1 − $\min_i$ (SE (x [i], y [i])))/False}, where x [i] and y [i] respectively denote the i-th components of x and y.

- Two bags x and y are equal if and only if each element that occurs in x occurs the same number of times in y. Meanwhile each element that occurs in y must also occurs in x. Equality $x = y$ is evaluated to be a fuzzy logical value {$\min_x$ ($\max_y$ SE (x [i], y [j]))/True, (1 − $\min_x$ ($\max_y$ SE (x [i], y [j])))/False}, where x [i] and y [j] respectively denote the i-th component of x and the j-th component of y. Note that i may be equal to j.

- The situation in which two lists x and y are equal is the same as that in arrays.

- The situation in which two sets x and y are equal is the same as that in bags.

Note that in the fuzzy aggregate data type, the boundary between fuzzy bag data type and fuzzy set data type becomes fuzzy because whether or not fuzzy elements are duplicates depends on the threshold. Therefore, in the comparison of fuzzy bags or fuzzy sets, it can be determined whether they are equal by measuring the equivalence degree of any two of their elements.

Entity Value Comparisons. The instance equal (:=:) and instance not equal (:<>:) operators produce a logical result. They operate only on compatible entity values. For two compatible entity values x and y, $x :=: y$ is evaluated to be {min (SE (x [A_i], y [A_i]))/True, (1 − min (SE (x [A_i], y [A_i])))/False}, where A_i is any explicit attribute declared in x and y (including attributes of all supertypes if any) and x [A_i] denotes the value of x on attribute A_i. It is clear that $x :<>: y$ is evaluated to be {(1 − min (SE (x [A_i], y [A_i])))/True, min (SE (x [A_i], y [A_i]))/False}.

Interval Expressions

An interval expression, having the form of $x\ \theta\ y\ \theta\ z$ where $\theta \in \{<, <=\}$, tests whether or not the value y falls within the given interval [x, z]. Here three operands must be compatible and have a defined ordering. It is permitted that operands be fuzzy.

The interval expression evaluates to a logical value. Since $x\ \theta\ y\ \theta\ z$ is equivalent to ($x\ \theta\ y$) AND ($y\ \theta\ z$), it can be evaluated through the value comparison operators above and the following logical operators together.

IN Operator

The IN operator tests an item for membership in some aggregate. Here, it is permitted that the right operand is a value of a fuzzy aggregate type and the left operand is compatible with the base type of the fuzzy aggregate value. So the IN operator may be fuzzy.

The expression x IN y is evaluated to a fuzzy logical value {max$_i$ (SE (x, y [i]))/True, (1 − max$_i$ (SE (x, y [i])))/False}, where y [i] denotes the i-th component of y.

LIKE Operator

The LIKE operator is a string-matching operator with the form of x LIKE y. It examines the target string x using the pattern string y as a control. Traditionally, the result is true if y matches x and false if the match fails.

Since these strings may be fuzzy strings and the pattern string may contain wildcards, the matching algorithm of the LIKE operator for fuzzy strings is very complex. Generally speaking, characters in the target string are compared to the corresponding character(s) of the pattern string. Only when all characters in the target match those in the pattern string (either exactly or as a wildcard) can the match succeed, and the answer is a fuzzy logical value. If any comparison does not match, then the result is false. For the former situation, the possibility of being true is the minimum of possibility degrees of all matched characters in the target and pattern strings. If there are several such matches, then the maximum is chosen. Note that if wildcards are used in the pattern string, the possibility degrees of the characters that the wildcards match are considered as 1.

Example 6.16: Let s_1 and s_2 be fuzzy strings, where s_1 = '{0.4/C, 0.5/A, 0.7/D, 0.9/D, 0.6/A, 0.3/T, 0.3/A}' and s_2 = '{0.2/C, 0.3/A, 0.4/D, 0.6/C, 0.7/A, 0.8/P, 0.8/P, 0.6/C, 0.5/A, 0.4/D}'.

s_1 LIKE 'CAD'	(*The result is {0.4/True, 0.6/False}*)
s_1 LIKE '{0.7/D, 0.5/A, ?, 0.2/A}'	(*The result is {0.2/True, 0.8/False}*)
s_2 LIKE 'CA'	(*The result is {0.6/True, 0.4/False}*)

Subset Operator

The subset operator (<=) accepts two compatible bags or sets and evaluates whether the first operand is a (not necessarily proper) subset of the second. The result may be fuzzy. Let a <= b.

Both a and b are set types. For $x \in a$ and $y \in b$, the possibility degree that a is a subset of b is

$$\min_x (\max_y SE (x, y)).$$

Then a <= b is evaluated to a fuzzy logical value {$\min_x$ ($\max_y$ SE(x, y))/True, $(1 - \min_x$ ($\max_y$ SE(x, y)))/False}.

Both a and b are bag types. For $x \in a$ and $y \in b$, the possibility degree that a is a subset of b is

$$\min_x (\max_y SE (x, y)).$$

Then a <= b is evaluated to a fuzzy logical value {$\min_x$ ($\max_y$ SE (x, y))/True, $(1 - \min_x$ ($\max_y$ SE (x, y)))/False}.

Example 6.17: Let s_1 and s_2 be fuzzy sets, where s_1 = [{0.8/2, 0.9/3, 0.4/4}, {0.4/5, 0.7/6}] and s_2 = [{0.2/1, 0.6/2, 0.8/3}, {0.7/2, 0.9/3, 0.4/4}, {0.3/6, 0.6/7, 0.9/8}].

s_1 <= s_2;	(*The result is {0.17/True, 0.83/False}*)

Superset Operator

The superset operator (>=) is the same as the subset operator (<=), only accepting two compatible bags and sets. Since a >= b can be regarded as b

<= a, the evaluation of subset operator can be similarly used for the super-set operator.

6.5.3 Logical Operators

Logical operators consist of NOT, AND, OR, and XOR. Each accepts one or two operands of type LOGICAL and produces a LOGICAL result. Considering fuzzy logical values, these four operators should be redefined. Let two fuzzy logical values be $a = \{\lambda_1/\text{True}, \zeta_1/\text{False}\}$ and $b = \{\lambda_2/\text{True}, \zeta_2/\text{False}\}$, respectively. Then

- NOT $a = \{(1 - \lambda1)/\text{True}, (1 - \zeta1)/\text{False}\}$
- a AND $b = \{\min(\lambda1, \lambda2)/\text{True}, \min(\zeta1, \zeta2)/\text{False}\}$
- a OR $b = \{\max(\lambda1, \lambda2)/\text{True}, \max(\zeta1, \zeta2)/\text{False}\}$
- a XOR $b = \{|\lambda1 - \lambda2|/\text{True}, (1 - |\zeta1 - \zeta2|)/\text{False}\}$

6.5.4 String and Binary Operators

String Indexing and Substring Indexing

Being similar to the classical situation, single components and components of a fuzzy string value can be addressed by a subscript and a pair of subscripts, respectively. Here, each subscript must be crisp. Note that each addressed component is connected with a membership degree that is the same as that of the corresponding component in the original string.

The methods for binary indexing and subbinary indexing are the same as that for string indexing and substring indexing.

Example 6.18: Let s be a fuzzy string and $s = $ '{0.7/C, 0.7/A, 0.7/D, 0.3/D, 0.3/A, 0.3/T, 0.3/A}'.

    x := s [3];   (*x contains the string '{0.7/D}'*)
    y := s [2 : 5];   (*y contains the string '{0.7/A, 0.7/D, 0.3/D, 0.3/A}'*)

String Concatenation

The string concatenation operator (+) is a binary operator that combines two strings together. This operator can be performed following the binary concatenations.

Example 6.19: Let s_1 and s_2 be fuzzy strings. Let $s_1 = $ '{0.5/C, 0.9/A, 0.7/D}' and $s_2 = $ '{0.3/C, 0.5/A, 0.7/S}'. Then

    z := s₁ + s₂;   (*z contains the string '{0.5/C, 0.9/A, 0.7/D, 0.3/C, 0.5/A, 0.7/S}'*)

6.5.5 Aggregate Operators

The aggregate operators are intersection (*), union (+), difference (−), and query. Since these operators are all concerned with duplicate elements in the operands, it is necessary to access if two elements, crisp or fuzzy, are redundant. The threshold β must be given for this purpose. The definition of semantic equivalence degree introduced in Section 3.1.2 is used again here. Let x and y be fuzzy values expressed by possibility distribution. It is said that x and y are duplicate with respect to β if SE $(x, y) \geq \beta$. If x or y is crisp, say x, it can be presented to a fuzzy value whose support is x and possibility is 1.

Intersection Operator

The intersection operator (*) accepts either two (fuzzy) sets or two (fuzzy) bags and evaluates to the same type of value. Each of its elements is the result of fuzzy intersection ($\cap_f$) of two elements from the operands, which are duplicates. The operator of fuzzy intersection has been defined in Chapter 3. Note that the operands must have compatible base types.

Example 6.20: Let s_1 and s_2 be sets of fuzzy integer data type and $s_1 = [\{0.5/1, 0.7/2\}, \{0.1/3, 0.6/4, 0.7/5\}, \{0.5/6, 0.9/7, 0.6/8\}]$ and $s_2 = [\{0.5/4, 0.7/5\}, \{0.4/7, 0.9/8, 0.3/9\}, \{0.4/10, 0.6/11\}]$. Let the threshold $\beta = 0.9$. Then $s_1 * s_2$ is evaluated to $[\{0.5/4, 0.7/5\}]$.

Union Operator

The union operator (+) evaluates to an aggregate value that is the combination of the elements in the first operand and those in the second. The following cases are classified:

- If the first operand is a set, then the second operand can be a set, element, bag, or list. Meanwhile, the result is also a set and the duplicate elements must be removed from the result by utilizing fuzzy union ($\cup_f$).
- If the first operand is a bag, then the second operand can be a set, element, bag, or list and the result is also a bag. The duplicate elements in the result cannot be considered.
- If the first operand is a list, then the second operand can be a list or an element and the result is also a list. The duplicate elements in the result cannot be considered.
- If the first operand is an element, then the second operand can be a list and the result is also a list. The duplicate elements in the result cannot be considered.

Example 6.21: Consider two sets s_1 and s_2 in example 6.20. Let the threshold $\beta = 0.9$. Then $s_1 + s_2$ is evaluated to $[\{0.5/1, 0.7/2\}, \{0.1/3, 0.6/4, 0.7/5\}, \{0.5/6, 0.9/7, 0.6/8\}, \{0.4/7, 0.9/8, 0.3/9\}, \{0.4/10, 0.6/11\}]$.

Difference Operator

The difference operator $(-)$ evaluates to an aggregate containing the elements in the second operand removed from the first. The type of the result will be the same as for the first operand. The following cases are classified:

- If the first operand is a set, then the second operand can be a set, element, or bag. The duplicate elements must be removed from the result by utilizing fuzzy difference $(-_f)$.
- If the first operand is a bag, then the second operand can be a set, element, or bag. If the first operand contains duplicate elements, only one of those elements is removed from each matching element in the second operand.

Example 6.22: Consider two sets s_1 and s_2 in example 6.20. Let the threshold $\beta = 0.9$. Then $s_1 - s_2$ is evaluated to $[\{0.5/1, 0.7/2\}, \{0.1/3, 0.1/4, 0.7/5\}, \{0.5/6, 0.9/7, 0.6/8\}]$.

Query Expression

The query expression evaluates a logical expression against each element of an aggregation, returning an aggregate containing only those elements for which the logical expression evaluates to true with respect to the given threshold. It is clear that the returned aggregate is a subset of the original aggregation where all of the elements of the subset satisfy the given condition.

Syntax: Query expression

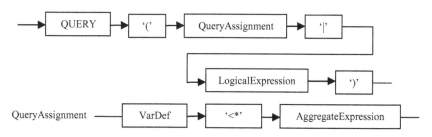

The first operand (*VarDef*) is implicitly declared as a local variable that exists only within the scope of this expression and is used in the *LogicalExpression* to hold the value being operated upon. The second operand

(*AggregateExpression*) is an aggregate value. The third operand (*Logica-lExpression*) is an expression that produces a fuzzy logical result.

Elements are first taken from the source aggregate and are placed in *VarDef* one by one. The logical expression is then evaluated. When the logical expression evaluates to true with respect to the given threshold, the value held by *VarDef* is added to the result; otherwise, it is not. The result aggregation is populated according to the specific kind of aggregation:

- Array. The result array has the same base type and bounds as the source set. The array elements are treated as optional values and each element is initially null. Any element in the source for which expression yields true is then placed at the corresponding index position in the result.
- Bag. The result bag has the same base type and upper bound as the source set. The lower bound is zero. The two cases correspond to the situations in which no element and all elements in the source aggregation satisfy the condition.
- List. The result list has the same base type and upper bound as the source set. The lower bound is zero. The two cases correspond to the situation in which no element and all elements in the source aggregation satisfy the condition. The result list is initially empty. Any element in the source for which expression yields true is then added to the end of the result. The order of the list is preserved.
- Set. The result set has the same base type and upper bound as the source set. The lower bound is zero. The two cases correspond to the situation in which no element and all elements in the source aggregation satisfy the condition.

6.6 Fuzzy Extensions to Interface and Executable Statements

6.6.1 The Interface Specification

Two specifications are used to establish an interface: USE and REFERENCE. Both USE and REFERENCE, which are stated immediately after the statement that gives the schema name if used, give access to declarations made in other schemas. The USE specification acts only upon entities, treating the foreign entity declaration as local. The REFERENCE specification treats the declarations as remaining remote but access is allow. As entities may be fuzzy at three levels, it is allowed that fuzzy entities occur in both USE and REFERENCE specifications.

6.6.2 Executable Statements

Executable statements define the actions of functions, procedures and rules. They define the logic and actions needed to support the definition of constraints by acting on parameters, local variables and constants. In this section, the executable statements concerned with fuzzy logic are considered.

Case Statement

The case statement consists of an expression, which is the case selector, and a list of alternative actions, each of which is preceded by a case label. Agreement between the type of the case label and the case selector is required. The first occurring statement having a case label that evaluates to the same value of the case selector is executed. At most, one case-action is executed. If none of the case labels evaluates to the same value as the case selector, the statement associated with otherwise is executed if it is present, or no statement associated with a case action is executed if it is not present.

According to the discussion above, it is known that the case selector may be evaluated to a fuzzy value. On the other hand, the case labels may be also provided fuzzily to implement flexible processing or the processing under incomplete knowledge environment. Therefore, it should be permitted to fuzzily evaluate if a case label has the same value of the case selector. Of course, such evaluation must be based on the given threshold. It is limited that the statement having a case label that has the highest equivalence degree with the value of the case selector is executed. If there are several such case labels, then the first occurring statement with such a case label is executed. If none of the case labels has a higher equivalence degree with the value of the case selector than the given threshold, the same processing as that in classical case statement is performed.

Syntax: Case statement

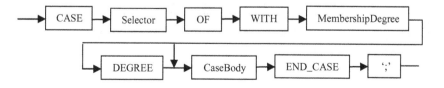

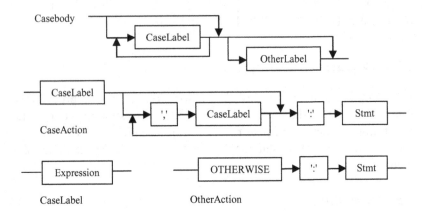

Casebody

CaseLabel

OtherLabel

CaseLabel

','   CaseLabel   ':'   Stmt

CaseAction

Expression

CaseLabel

OTHERWISE   ':'   Stmt

OtherAction

Example 6.23: This case statement executes one several statements depending on the value of a selector.

```
LOCAL
      a : FUZZY INTEGER;
      x : REAL;
END_LOCAL;

      ...
      a := {0.1/4, 0.5/5, 0.7/6, 0.4/7};
      x := 34.97;
      CASE a OF WITH 0.90 DEGREE
            {0.4/1, 0.7/2, 0.5/3, 0.3/4} : x := SIN (x);
            {0.1/3, 0.3/4, 0.6/5, 0.8/6} : x := EXP (x);
            {0.5/5, 0.7/6, 0.4/7, 0.1/8} : x := LOG (x); (*This is executed*)
            OTHERWISE : x := 0.0;
      END_CASE;
```

If ... Then ... Else Statement

The IF ... THEN ... ELSE statement allows the conditional execution of statements based on the value of a logical expression. When the logical expression evaluates to true with respect to the given threshold, the statement after Then is executed. When the logical expression evaluates false with respect to the given threshold, the statement after Else is executed if else phrase is resent. If the Else phrase is omitted meanwhile, the control is passed to the next statement. Here, the conditional expression may be evaluated to a fuzzy logical value. The given threshold is used to determine which statement is to be executed.

Syntax: If ... then ... else statement

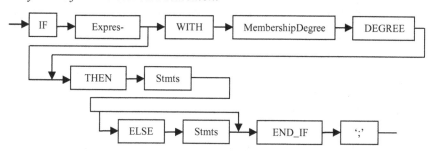

Repeat Statement

The repeat statement is used to execute a sequence of statements a number of times based on the outcome of various control conditions. Those control conditions are finite iteration, while a condition is true, and until a condition is true.

Increment control. For this case, the value of the increment control variable is incremented by step. If the iteration control variable is between the given lower bound and the upper bound, the body of the repeat statement is executed; otherwise the execution of the repeat statement is terminated. The step and the two bounds may be expressions. The values of these expressions may be evaluated fuzzy. In this situation, the execution of the repeat statement can be controlled by means of fuzzy logic, fuzzy calculation and the given threshold, but it is generally limited that the repeat statement is not executed if any of the expressions have no a crisp value. This is the same as in classical increment control. As to the repeat statement under fuzzy condition, while control and until control can be used.

While control. The While control continues the execution of the body of the repeat statement while the control expression is true with respect to the given threshold. The expression is evaluated before each iteration. If the While control expression evaluates to false initially, the body will not be executed.

Syntax: While control

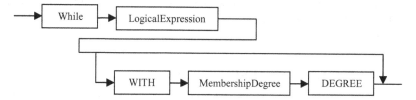

Until control. The Until control continues the execution of the body of the repeat statement until the control expression evaluates to true with respect to the given threshold. The expression is evaluated after each iteration. If the Until control expression is only control present, at least one iteration will always be executed.

Syntax: Until control

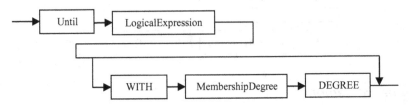

6.7 The Fuzzy EXPRESS-G Model

EXPRESS-G is the graphical representation of EXPRESS, which uses graphical symbols to form a diagram. Note that EXPRESS-G can only express a subset of the full language of EXPRESS. EXPESS-G provides supports for the notions of entity, type, relationship, cardinality, and schema. The functions, procedures, and rules in EXPRESS language are not supported by EXPRESS-G. In the following, EXPRESS-G is briefly reviewed.

Since entity and schema are keys of EXPRESS-G information model, in the following, we extend EXPRESS-G for fuzzy information modeling in entity level and schema level models, respectively. In entity level model, we mainly investigate the fuzziness in data types, attributes, entities, and relationships. In schema level, we mainly investigate the fuzziness between schemas. The corresponding notations are hereby introduced.

6.7.1 Fuzziness in Entity Level Model

An entity-level model is an EXPRESS-G model that represents the definitions and relationships that comprise a single schema. So the components of such a model consist of type, entity, relationship symbol, role, and cardinality information.

Imprecise and Uncertain Role Names

As we know, entities consist of attributes and an attribute corresponds to a domain of values with a certain type. In EXPRESS-G, the attributes of an entity are role named. The text string representing the role name is placed on the relationship line connecting an entity symbol to its attribute representation. For an attribute of aggregation, the role name on the relationship line is followed by the first letter of the aggregation such as *LIST* [*m: n*], *ARRAY* [*m: n*], *SET* [*m: n*], or *BAG* [*m: n*].

It should be noted that, however, the attribute values of an entity may be imprecise or fuzzy. An entity *Engine*, for example, has an attribute *Size*, which data type is *Real* and values might be either fuzzy or imprecise in preliminary phase of its design. Such role names that may take imprecise and fuzzy values are different from one that only takes crisp values. In fact, EXPRESS has considered the situation that attributes can take imprecise null values using the keyword "*optional*" in explicit attribute definition. Modeling and identifying the role names taking imprecise range values or uncertain fuzzy values, however, have not yet been discussed in EXPRESS. The symbols for range-valued and fuzzy attributes which take the values of simple data types are shown in Figure 6.1 and Figure 6.2, respectively.

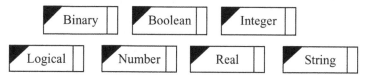

Fig. 6. 1. Simple range-valued type symbols in the fuzzy EXPRESS-G

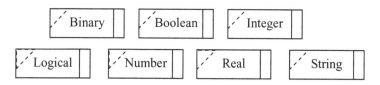

Fig. 6.2. Simple fuzzy type symbols in the fuzzy EXPRESS-G

Imprecise and Uncertain type Modeling

Also fuzziness can be found in type modeling. First let's have a look at enumeration type. As we know, an enumeration type is an ordered list of values represented by name, where the list has a perfect boundary. A value named either belongs to the enumeration type or does not belong to the enumeration type. It is possible, however, that a value belongs to the enumeration type with degree, namely, the value is fuzzy. For example, an enumeration type is *HairType* = *ENUMERATION OF* (*Black, Red, Brown, Golden, Grey*) and the hair type of a person is *red and brown*.

A defined data type is created based on the underlying type. The defined data type generally has the same domain of values as the underlying type unless a constraint is put on it. The underlying type can be simple type, collection type, enumeration type, select type, and named type. It has been shown that a value of simple type, collect type, or enumeration type may be fuzzy or imprecise. The imprecision and fuzziness for the values of select type or entity type are shown in the following. Thus, the value of a defined data type may be imprecise or fuzzy.

A select type defines a named collection of other types called a select list. A value of a select type is a value of one of the types specified in the select list where each item is an entity type or a defined type. The imprecision or fuzziness of a value of select type comes from the imprecision or fuzziness of its component type, fuzzy or imprecise entity type and defined type.

Fig. 6.3. Range-valued type definition symbols in the fuzzy EXPRESS-G

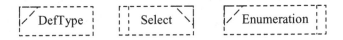

Fig. 6.4. Fuzzy type definition symbols in the fuzzy EXPRESS-G

The symbols for modeling imprecise and fuzzy type are shown in Figure 6.3 and Figure 6.4, respectively.

Fuzzy Entity Modeling

Fuzzy entity modeling can be classified into two levels. The first level is the fuzziness in the entity sets, i.e., an entity has degree of membership in the model. For example, an entity *Engine* may be fuzzy in the product data model. The second level is related to the fuzzy occurrences of entities. For an entity *Research Student*, for example, it is not uncertain if John is a Ph.D. student. Such an entity is represented using the definition symbol in Figure 6.5.

For the first level fuzziness, memberships can be placed inside the solid rectangle as well as the name of the entities. Let E be an entity and $\mu(E)$ be its grade of membership in the model, then "$\mu(E)/E$" is enclosed in the solid rectangle. If $\mu(E) = 1.0$, "$1.0/E$" is usually denoted "E" simply. The graphical representation of such entity is shown in Figure 6.6.

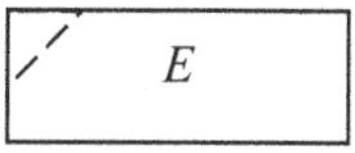

Fig. 6.5. Entity with fuzzy instances in EXPRESS-G

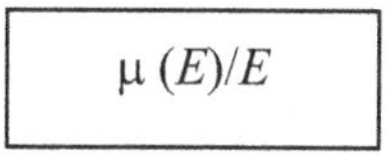

Fig. 6.6. Entity with membership degree in EXPRESS-G

In classical situation, if there exist two entities E_1 and E_2 such that for any entity instance e, e $\in E_2$ implies e $\in E_1$, then E_2 is called a subtype of E_1, and E_1 is called a supertype of E_2. As mentioned above, an instance of entity, say e, may be fuzzy for an entity, say E. Therefore, there exists fuzzy supertype/subtype in EXPRESS. Let E and S be two fuzzy entities with membership functions μ_E and μ_S, respectively. Then S is a fuzzy subtype of E and E is a fuzzy supertype of S if and only if the following is true.

$$(\forall e)\,(e \in U \wedge \mu_S(e) \le \mu_E(e))$$

Considering a fuzzy supertype E and its fuzzy subtypes $S_1, S_2, ..., S_n$ with membership functions $\mu_E, \mu_{S1}, \mu_{S2, ...}$, and μ_{Sn}, respectively, the following relationship is true.

$$(\forall e)\,(\forall S)\,(e \in U \wedge S \in \{S_1, S_2, ..., S_n\} \wedge \mu_S(e) \le \mu_E(e))$$

For the fuzzy subtype with multiple fuzzy supertypes, let E be a fuzzy subtype and $S_1, S_2, ..., S_n$ be its fuzzy supertypes, which membership functions are respectively $\mu_E, \mu_{S1}, \mu_{S2, ...}$, and μ_{Sn}.

$$(\forall e)\,(\forall S)\,(e \in E \wedge S \in \{S_1, S_2, ..., S_n\} \wedge \mu_E(e) > 0 \wedge \mu_S(e) \ge \mu_E(e))$$

The fuzzy supertype/subtype in fuzzy EXPRESS-G can be represented with fuzzy relationship.

Fuzzy Relationship Modeling

As mentioned above, there are dashed lines, thick solid lines, and thin solid lines in EXPRESS-G. Dashed lines and thin solid lines connecting attributes represents that attribute must belong to the corresponding entity. It is possible that an attribute is a fuzzy one with membership function. So memberships should be placed upon dashed lines and thin solid lines. In addition, dashed thick solid lines are used to represent fuzzy super-type/subtype above.

The symbols for these three lines are shown in Figure 6.7, where A and μ denote the name of an attribute and its degree of membership, respectively.

μ (A)/A   μ (A)/A

Fig. 6.7. Styles of the fuzzy relationship lines in EXPRESS-G

6.7.2 Fuzziness in Schema Level Model

A schema-level model is one that displays the schemas, and the relationships between these schemas. A schema may be fuzzy because it consists of fuzzy entities or fuzzy relationship. The relationships between these fuzzy schemas are thus fuzzy. The *Use* and *Reference* relationships in fuzzy EXPRESS-G, which mean the fuzzy relationships between fuzzy schemas, are denoted normal width relation lines and dashed relation lines with memberships. The symbol for a fuzzy schema is shown in Figure 6.8.

Through the discussion above, three levels of fuzziness can bee found in fuzzy EXPRESS-G, namely, the fuzziness at the level of attribute value (the third level), the fuzziness at the level of instance/entity (the second level), and the fuzziness at the level of entity and attribute (the first level). The fuzziness at the third level means that attributes take fuzzy values. The second level of fuzziness means that each instance of an entity belongs to the entity with a membership degree. The first level of fuzziness means that attributes comprise an entity with membership degrees or entities comprise a schema with membership degrees.

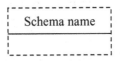

Fig. 6.8. The fuzzy schema definition symbol in EXPRESS-G

6.8 Applicability of the Fuzzy EXPRESS

The information modeling process with EXPRESS is to describe or present the information concerning the *data types, declarations, expressions, executable statements,* and *interfacing.* Under classical environment, the classical EXPRESS can be used to create EXPRESS data model. But the classical EXPRESS suffers from its incapability of representing and manipulating imprecise and uncertain information that may occur in many engineering activities. The fuzzy EXPRESS is hereby proposed and applied for fuzzy engineering information modeling.

Being the same as the classical EXPRESS, the fuzzy EXPRESS also needs to describe the fuzzy information concerning the data types, declarations, expressions, executable statements, and interfacing as the extension to the classical EXPRESS. Basically we can classify the constructs of the fuzzy EXPRESS for fuzzy information modeling into two kinds: one is for the representation of fuzzy information (including fuzzy data types and declarations) and another is for the processing of fuzzy information (including fuzzy expressions, executable statements, and interfacing). What the later kind of constructs process is the objects described by the former kinds of constructs. There exist two reasons why fuzzy expressions, executable statements, and interfacing are proposed. The first one is that the objects described by the former kinds of constructs contain fuzzy information and we need the fuzzy expressions, executable statements, and interfacing to deal with them. As a matter of fact, we may also need the fuzzy expressions, executable statements, and interfacing for flexible, intelligent, and/or robust information processing (e.g., flexible information query) even if the objects described by data types and declarations do not contain any imperfect information. But we must apply fuzzy data types and declarations to describe the objects we create to represent information of interest to us if they contain fuzzy information.

It is not difficult to find that one of the core tasks of fuzzy information modeling with EXPRESS is to create a fuzzy information model (i.e., fuzzy EXPRESS data model) containing the declarations and fuzzy data types. It should be noticed that the fuzzy EXPRESS data model is generally designed starting from fuzzy EXPRESS data model. That is, we first create a fuzzy EXPRESS-G data model using the fuzzy EXPRESS-G and then this EXPRESS-G data model is converted into the fuzzy EXPRESS data model. In the following, we present a simple product example containing fuzzy information and give its EXPRESS-G data model and EXPRESS data model.

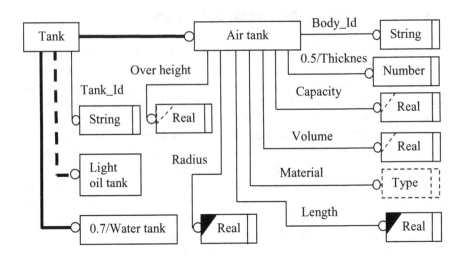

Fig. 6.9. A fuzzy EXPRESS-G data model

Assume we have the following product example. Entity *Tank* is a super-type, which has three subtypes, namely, *Air tank, Water tank,* and *Light oil tank.* Among these three subtypes, it is known for certain that entities *Light oil tank* and *Air tank* are the subtypes of entity *Tank.* In other word, the membership degrees that *Light oil tank* and *Air tank* are the subtypes of *Tank* are 1.0, respectively. However, it is not known for certain if entity *Water tank* must be the subtype of entity *Tank.* It is only known that the membership degree that *Water tank* is the subtypes of *Tank* is 0.7. In addition, entity *Light oil tank* is a fuzzy entity with fuzzy instances. The entity *Air tank* has eight attributes. The attribute *Body_Id* is a perfect one with string type. Attribute *Thickness* is an attribute associated with a member-ship degree 0.5, which means the possibility that entity *Air tank* has attribute *Thickness* is not certainly 1.0 but is only 0.5. The attributes *Volume, Capacity, Over height, Length,* and *Radius* are ones that can take fuzzy real values. It should be noted that attribute *Material* is one of enumeration type.

Utilizing some notations introduced in this section, we create a fuzzy EXPRESS-G data model for the product example above as shown in Fig-ure 6.9. The EXPRESS data model of the fuzzy EXPRESS-G data model in Figure 6.9 is given in Figure 6.10.

```
SCHEMA
    ENTITY Tank
        Tank_Id: String;
    END_ENTITY;
    ENTITY Air tank
```

```
    SUBTYPE OF Tank;
        Body_Id: String;
        Thickness: Number WITH 0.5 DEGREE;
        Capacity: FUZZY Real;
        Volume: FUZZY Real;
        Length: FUZZY Real;
        Over height: FUZZY Real;
        Radius: FUZZY Real;
        Material: FUZZY Type;
    END_ENTITY;
    TYPE Type = ENUMERATION (...);
    END_TYPE;
    ENTITY Water tank WITH 0.7 DEGREE
    SUBTYPE OF Tank;
    END_ENTITY;
    ENTITY Light oil tank
    SUBTYPE OF Tank;
        pD: NUMBER;
    END_ENTITY;
END_SCHEMA
```

Fig. 6.10. A fuzzy EXPRESS data model

6.9 Conceptual Design of the Fuzzy EXPRESS Model

In contrast to ER/EER and IDEF1X, EXPRESS is not a graphical schema language. In order to construct EXPRESS data model at a higher level of abstract, EXPRESS-G is introduced as the graphical representation of EXPRESS. Here EXPRESS-G can only express a subset of the full language of EXPRESS, which provides supports for the notions of entity, type, relationship, cardinality, and schema. The functions, procedures, and rules in EXPRESS language are not supported by EXPRESS-G. In addition to EXPRESS-G, it is also suggested in STEP that IDEF1X or ER/EER can be used as one of the optional languages for EXPRESS data model design. Then EXPRESS-G, IDEF1X, ER/EER, or even UML data model can be translated into EXPRESS data model. That multiple graphical data models can be employed facilitates the designers with different background to design their EXPRESS models easily by using one of the graphical data models that they are familiar with. There are already some efforts for converting EXPRESS-G, IDEF1X, ER/EER, or UML data model into EXPRESS data model.

The conceptual design of the fuzzy EXPRESS model can follow the same way. However, a complex EXPRESS data model is generally completed cooperatively by a design group, in which each member may use a different graphical data model (Ma, Lu and Fotouhi, 2003). All these graphical data models designed by different members should be converted into one consistent data model in order to create EXPRESS data models. Then EXPRESS-G is chosen as the target data model and the other graphical data models should be converted into EXPRESS-G. Originated from EXPRESS, it is not difficult to convert EXPRESS-G model into EXPRESS model. It is also true for the mapping of the fuzzy EXPRESS model from the fuzzy EXPRESS-G. But the data model conversions from IDEF1X, ER/EER and UML to EXPRESS-G only receive few attentions although such conversions are crucial in engineering information modeling. EER and EXPRESS-G models are known well and popular in the areas of database design and engineering information modeling, respectively. In the following, we investigate the basic issue about the formal transformation of EER and EXPRESS-G.

To implement the formal transformation of EER and EXPRESS-G, it is necessary to compare their capabilities in data modeling and investigate how they match each other. First of all, we introduce the constructs in each of these two data models briefly. In addition to *entity*, *relationship*, and *attribute*, the following notions are introduced in the EER: *specialization/generalization*, *category*, and *aggregation*. EXPRESS-G provides supports for the notions of *entity*, *type*, *relationship*, *cardinality*, and *schema*. The notions *entity* in EER and *entity* in EXPRESS-G mean the same thing. There are three kinds of *relationship* in EXPRESS-G: *optional valued attributes of an entity*, *supertype/subtype relationships*, and *common attributes of an entity*, which are denoted by *dashed lines*, *thick solid lines*, and *thin solid lines*, respectively. The notion *attribute* in EER corresponds to *dashed lines* and *thin solid lines* in EXPRESS-G. The notion *specialization/generalization* in EER corresponds to *thick solid lines* in EXPRESS-G. In addition, the notion *relationship* in EER can be regarded as the *entity* in EXPRESS-G. Now let us focus on some incompatibilities between EER and EXPRESS-G. The notions *category* and *aggregation* are only supported by EER and the notions *type* and *schema* are only supported by EXPRESS-G.

For the compatibilities between EER and EXPRESS-G, we can conduct mapping from EER to EXPRESS-G and ever from EXPRESS-G to EER. As to the incompatibilities between EER and EXPRESS-G, we can simulate *category* and *aggregation* of EER in EXPRESS-G. But current EER cannot support *type* and *schema*. In order to model *type* and *schema*, EER has to be extended.

Based on the discussions above, we can draw some strategies for the conceptual design of the fuzzy EXPRESS model. First, we can use the fuzzy EXPRESS-G to conceptually design the fuzzy EXPRESS model.

Second, we can directly convert the compatibilities between EER and EXPRESS-G from the fuzzy EER model into the fuzzy EXPRESS-G model. Then the incompatibilities between EER and EXPRESS-G can be added into the converted fuzzy EXPRESS-G model directly.

6.10   Summary

Based on fuzzy set and fuzzy logic, an original and full extension of EXPRESS language on fuzzy information modeling has been developed in this chapter. Note that in this chapter, only the extended parts of the current edition of EXPRESS, including basic elements, various data types, EXPRESS declarations, calculation and operations, and EXPRESS-G, are addressed. Other parts that are not mentioned can directly be used in the extended EXPRESS because they are irrelevant to imprecise and uncertain information modeling. In addition, the extended parts fully cover the corresponding parts in the current edition of EXPRESS. Therefore, utilizing the extended EXPRESS, the information model with imperfect information as well as crisp information can be constructed. When there is no any imprecision and uncertainty in information, the extended EXPRESS resumes to the current edition of EXPRESS.

EXPRESS data modeling language is used to describe a product data model with activities covering the whole product life cycle. Based on such a product data model, product data can be exchanged and shared among different applications. This is the goal of the STEP standard. Generally speaking, the application of STEP is mainly concerned with two aspects. One of the aspects is the establishment of the product information model to represent product data according to information requirements in application environment and the integrated resources in STEP. The other one is the manipulation and management of product data in the product information model. All these are related to the implementation of STEP in database systems (Loffredo, 1998). The issues on how to implement the fuzzy EXPRESS data model in database systems and reach the goal of STEP will be stated in Chapter 9 and Chapter 10.

References

ISO IS 10303-1 TC184/SC4 (1994), Product Data Representation and Exchange-Part 1: Overview and Fundamental Principles, International Standard.

ISO IS 10303-1 TC184/SC4 (1994), Product Data Representation and Exchange-Part 11: The EXPRESS Language Reference Manual, International Standard.

Loffredo, D. (1998), Efficient Database Implementation of EXPRESS Information Models, Ph.D. Thesis, Rensselaer Polytechnic Institute, Troy, New York.

Ma, Z. M. (2004), Fuzzy declaration modeling of data models using EXPRESS, Proceedings of the 2004 ASME Design Engineering Technical Conference and Computers and Information in Engineering Conference, 4: 45-51.

Ma, Z. M. (2005), Expressions with fuzzy information in EXPRESS/STEP, Proceedings of the 2005 ASME Computers and Information in Engineering Conference (CD).

Ma, Z. M., Zhang, W. J. and Ma, W. Y. (2001), Fuzzy data type modeling with EXPRES, Proceedings of the 2001 ASME Design Engineering Technical Conference and Computers and Information in Engineering Conference, 1: 99-105.

Ma, Z. M., Zhang, W. J., Ma, W. Y. and Chen, G. Q. (2000), Extending EXPRESS-G to model fuzzy information in product data model, Proceedings of the ASME 2000 International Design Engineering Technical Conferences and the Computers and Information in Engineering Conference (CD).

Schenck, D. A. and Wilson, P. R. (1994), Information Modeling: the EXPRESS Way, Oxford University Press.

Zadeh, L. A. (1975), The concept of a linguistic variable and its application to approximate reasoning, Information Sciences, 8: 119-249 & 301-357; 9: 43-80.

Zimmermann, H. J. (2001), Fuzzy Set Theory and Its Applications (Fourth Edition), Kluwer Academic Publishers.

7 The Fuzzy Logical Databases

7.1 Introduction

A major goal for database research has been the corporation of additional semantics into the database model. Classical database models often suffer from their incapability of representing and manipulating imprecise and uncertain information that may occur in many real world applications. Since the early 1980's, Zadeh's fuzzy logic has been used to extend various database models. The purpose of introducing fuzzy logic in databases is to enhance the classical database models such that uncertain and imprecise information can be represented and manipulated. This resulted in numerous contributions, mainly with respect to the popular relational model or to some related form of it.

However, rapid advances in computing power have brought opportunities for databases in emerging applications in CAD/CAM, multimedia and geographic information systems (GIS). These applications characteristically require the modeling and manipulation of complex objects and semantic relationships. Since classical relational database model and its extension of fuzziness do not satisfy the need of modeling complex objects with imprecision and uncertainty, currently many researches have been concentrated on fuzzy nested relational and object-oriented database models in order to deal with complex objects and uncertain data together.

7.2 The Fuzzy Relational Databases

Fuzzy data have been used to model imprecise information in databases since Zadeh introduced the concept of fuzzy sets (Zadeh, 1965), and the traditional relational databases have been thereby extended. Much of the work in the area has been in extending the basic data model and query language to permit the representation and retrieval of imprecise data. A

Z. Ma: *Fuzzy Database Modeling of Imprecise and Uncertain Engineering Information*,
StudFuzz **195**, 137–158 (2006)
www.springerlink.com

number of related issues such as data dependencies, implementation considerations and others have also been investigated in various fuzzy relational database models.

7.2.1 The Fuzzy Relational Database Models

Fuzzy data is originally described as fuzzy set (Zadeh, 1965) or equivalently possibility distribution (Zadeh, 1978). In addition, a fuzzy data can also be represented by similarity relations in domain elements (Buckles and Petry, 1982), in which the fuzziness comes from the similarity relations between two values in a universe of discourse, not from the status of an object itself. Similarity relations are thus used to describe the similarity degree of two values from the same universe of discourse. A similarity relation Sim on the universe of discourse U is a mapping: $U \times U \rightarrow [0, 1]$ such that

- for $\forall x \in U$, $Sim(x, x) = 1$,   (reflexivity)
- for $\forall x, y \in U$, $Sim(x, y) = Sim_i(y, x)$, and   (symmetry)
- for $\forall x, y, z \in U$, $Sim(x, z) \geq \max_y(\min(Sim(x, y), Sim(y, z)))$.
                                                                   (transitivity)

Based on the similarity relations, the proximity relations (Shenoi and Melton, 1989) and the resemblance relations (Rundensteiner et al., 1989) are further proposed. In connection to the three types of fuzzy data representations, there exist two basic extended data models for fuzzy relational databases. The first one of the fuzzy relational models is based on similarity relations (Buckles and Petry, 1982), or proximity relations (Shenoi and Melton, 1989), or resemblance relations (Rundensteiner et al., 1989). The form of an n-tuple in the similarity-based fuzzy relational model can be expressed as

$$t = <p_1, p_2, ..., p_i, ..., p_n>,$$

where $p_i \subseteq D_i$ with D_i being the domain of attribute A_i. Viewed from expressive format, p_i is a set value containing some elements of D_i, in which these elements have similarity relations. In other words, p_i is a fuzzy data represented by similarity relations in the elements of D_i.

The second one of the fuzzy relational models is based on possibility distribution (Prade and Testemale, 1984; Raju and Majumdar, 1988), which can be further classified into two basic categories. The first one is that tuples are associated with possibility degrees but attribute values are crisp (*type 1*) (Raju and Majumdar, 1988) and the second one is that attribute values are fuzzy ones represented by possibility distributions or similarity relationships (*type 2*) (Buckles and Petry, 1982; Prade and

Testemale, 1984). According to (Raju and Majumdar, 1988), in a type 1 fuzzy relation, the attribute domain dom (A_i) may be a classical subset or a fuzzy subset of the universe of discourse U_i. Let the membership function of dom (A_i) be denoted by μ_{Ai} $(i = 1, ..., n)$ and $U = U_1 \times U_2 \times ... \times U_n$. A type 1 fuzzy relation r is hereby a fuzzy subset of U with membership function μ_r, where $\mu_r (u_1 u_2...u_n) \leq \min (\mu_{A1} (u_1), \mu_{A2} (u_2), ..., \mu_{An} (u_n))$ for all $(u_1 u_2...u_n)$ $\in U$. According to possibilistic interpretation of fuzzy sets, μ_r can be treated as a possibility distribution function in U. Thus $\mu_r (u_1 u_2...u_n)$ determines the possibility that a tuple $t \in U$ has t $[A_i] = u_i$ $(i = 1, ..., n)$. In other words, $\mu_r (u_1 u_2...u_n)$ is a fuzzy measure of association among a set of domain values $\{u_1, u_2, ... , u_n\}$. Combining these two types of fuzzy relational model, we have the fuzzy relational model of *type 3* (Raju and Majumdar, 1988), where attribute values may be fuzzy and tuples are associated with possibility degrees. It is clear that the fuzzy relational model of type 3 is a hybrid of the fuzzy relational models type 1 of and type 2. The type 3 fuzzy relation was called *fuzzy-2* fuzzy relation in (Raju and Majumdar, 1988). Here for any attribute A_i, dom (A_i) may be a set of fuzzy sets in U_i. A tuple $t = <a_1, a_2, ..., a_n>$ in $D = \text{dom} (A_1) \times \text{dom} (A_2) \times ... \times \text{dom} (A_n)$ becomes a fuzzy subset of $U = U_1 \times U_2 \times ... \times U_n$, with $\mu_t (u_1 u_2...u_n) = \min (\mu_{a1} (u_1), \mu_{a2} (u_2), ..., \mu_{an} (u_n))$, where $u_i \in U_i$ $(i = 1, ..., n)$. A fuzzy relation r is hereby a fuzzy subset of D and the membership function $\mu_r: D \rightarrow [0, 1]$ satisfies $\mu_r (t) = \max_{(U1 \times U2 \times ... \times Un) \in U} (\min (\mu_{a1} (u_1), \mu_{a2} (u_2), ..., \mu_{an} (u_n)))$, where $t = <a_1, a_2, ..., a_n> \in D$.

The form of an n-tuple for each of the possibility-based fuzzy relational models can be expressed, respectively, as

$$t = <a_1, a_2, ..., a_i, ..., a_n, pD > \text{ for type 1,}$$

$$t = <\pi_{A1}, \pi_{A2}, ..., \pi_{Ai}, ..., \pi_{An}> \text{ for type 2, and}$$

$$t = <\pi_{A1}, \pi_{A2}, ..., \pi_{Ai}, ..., \pi_{An}, pD > \text{ for type 3.}$$

Here, $a_i \in D_i$ with D_i being the domain of attribute A_i, $pD \in (0, 1]$, π_{Ai} is the possibility distribution of attribute A_i on its domain D_i, and $\pi_{Ai} (x)$, $x \in D_i$, denotes the possibility that x is true. In the fuzzy relation in type 1 and type 3, attribute pD is used to denote the membership value of the tuples in the relation. In other words, the values of attribute pD of tuples may be interpreted either as a possibility measure of association among the data values or as a truth value of a fuzzy predicate associated with the relation. In this paper, we focus on the last type of fuzzy relational databases. It should be pointed out that, however, the fuzzy relational model of type 2 can be viewed as the special case of the fuzzy relational model of type 3. When the membership degrees of all tuples in the fuzzy relational model of type

3 are 1.0, attribute *pD* can be omitted and the fuzzy relational model of type 3 changes into the fuzzy relational model of type 2.

Based on the above-mentioned similarity-based and possibility-based fuzzy relational models, it is clear that one can combine possibility distribution and similarity (proximity or resemblance) relation. The *extended possibility-based fuzzy relational databases* are hereby proposed in (Chen et al., 1992; Ma et al. 2000; Rundensteiner et al., 1989), where possibility distribution and resemblance relation arise in a relational database simultaneously.

Definition: A fuzzy relation *r* on a relational schema *R* (A_1, A_2, ..., A_n) is a subset of the Cartesian product of Dom (A_1) × Dom (A_2) × ... × Dom (A_n), where Dom (A_i) may be a fuzzy subset or even a set of fuzzy subset and there is the resemblance relation on the Dom (A_i). A resemblance relation *Res* on Dom (A_i) is a mapping: Dom (A_i) × Dom (A_i) → [0, 1] such that

- for all x in Dom (A_i), *Res* (x, x) = 1   (reflexivity)
- for all x, y in Dom (A_i), *Res* (x, y) = *Res* (y, x)   (symmetry)

Based on various fuzzy relational database models, many studies have been done for data integrity constraints (Bosc and Pivert, 2003; Bosc et al., 1998; Sözat and Yazici, 2001; Raju and Majumdar, 1988; Liu, 1997). Also there have been research studies on fuzzy query languages (Bosc and Pivert, 1995; Takahashi, 1993) and fuzzy relational algebra (Umano and Fukami, 1994; Ma and Mili, 2002b). In (Zemankova and Kandel, 1985), the fuzzy relational data base (FRDB) model architecture and query language were presented and the possible applications of the FRDB in imprecise information processing were discussed. For a comprehensive review of what has been done in the development of the fuzzy relational databases, please refer to (Petry, 1996; Chen, 1999; Yazici and George, 1999; Ma, 2005; Yazici et al., 1992).

Among the issues investigated in the fuzzy relational databases, the semantic measure of fuzzy data is related to dealing with fuzzy information and is hereby crucial.

7.2.2 Semantic Measure and Data Redundancies

To measure the semantic relationship between fuzzy data, some investigation results for assessing data redundancy can be found in literature.

(a) Rundensteiner et al. in (1989) proposed the notion of nearness measure. Two fuzzy data π_A and π_B were considered α-β redundant if and only if the following inequality equations hold true:

$\min_{x, y \in \text{supp}(\pi A) \cup \text{supp}(\pi B)} (\text{Res}(x, y)) \geq \alpha$ and $\min_{z \in U} (1 - |\pi_A(z) - \pi_B(z)|)$
$$\geq \beta,$$

where α and β are the given thresholds, $\text{Res}(x, y)$ denotes the resemblance relation on the attribute domain, and $\text{supp}(\pi_A)$ denotes the support of π_A. It is clear that a twofold condition is applied in their study.

(b) For two data π_A and π_B, Chen $et\ al.$ in (1992) defined the following approach to assess the possibility and impossibility that $\pi_A = \pi_B$.

$$E_c(\pi_A, \pi_B)(T) = \text{supp}_{x, y \in U, c(x, y) \geq \alpha} (\min(\pi_A(x), \pi_B(y))) \text{ and}$$

$$E_c(\pi_A, \pi_B)(F) = \text{supp}_{x, y \in U, c(x, y) < \alpha} (\min(\pi_A(x), \pi_B(y)))$$

Here $c(x, y)$ denotes a closeness relation (being the same as the resemblance relation).

(c) In (Cubero and Vila, 1994), the notions of weak resemblance and strong resemblance were proposed for representing the possibility and the necessity that two fuzzy values π_A and π_B are approximately equal, respectively. Weak resemblance and strong resemblance can be expressed as follows.

$$\Pi(\pi_A \approx \pi_B) = \text{supp}_{x, y \in U} (\min(\text{Res}(x, y), \pi_A(x), \pi_B(y))) \text{ and}$$

$$N(\pi_A \approx \pi_B) = \inf_{x, y \in U} (\max(\text{Res}(x, y), 1 - \pi_A(x), 1 - \pi_B(y)))$$

The semantic measures were employed as a basis for a new definition of fuzzy functional dependencies in (Cubero and Vila, 1994).

(d) Bosc and Pivert (1997) gave the following function to measure the interchangeability that fuzzy value π_A can be replaced with another fuzzy data π_B, i.e., the possibility that π_A is close to π_B from the left-hand side:

$$\mu_{\text{repl}}(\pi_A, \pi_B) = \inf_{x \in \text{supp}(\pi A)} (\max(1 - \pi_A(x), \mu_S(\pi_A, \pi_B)(x))),$$

where $\mu_S(\pi_A, \pi_B)(x)$ is defined as

$$\mu_S(\pi_A, \pi_B)(x) = \sup_{y \in \text{supp}(\pi B)} (\min(\text{Res}(x, y), 1 - |\pi_A(x) - \pi_B(y)|)).$$

It has been shown that counterintuitive results are produced with the treatment of (a) due to the fact that two criteria are set separately for redundancy evaluation (Chen $et\ al.$, 1992; Bosc and Pivert, 1997). Therefore the approaches of (b) and (d) tried to set two criteria together for the redundancy evaluation. But for the approach in (b), there also exist some inconsistencies for assessing the redundancy of fuzzy data represented possibility distribution (Ma $et\ al.$, 1999). The approach in (d) is actually an extension of the approach of (a) and the counterintuitive problem in (a) still exists in the approach in (d), which has been demonstrated in

(Ma *et al.* 2000). As to the approach in (c), the weak resemblance, however, appears to be too "optimistic" and strong resemblance is too severe for the semantic assessment of fuzzy data (Bosc and Pivert, 1997). So in (Ma *et al.* 2000), the notions of semantic inclusion degree and semantic equivalence degree were proposed.

For two fuzzy data π_A and π_B, semantic inclusion degree SID (π_A, π_B) denotes the degree that π_A semantically includes π_B and semantic equivalence degree SED (π_A, π_B) denote the degree that π_A and π_B are equivalent to each other. Based on possibility distribution and resemblance relation, the definitions of calculating the semantic inclusion degree and the semantic equivalence degree of two fuzzy data are given as follows.

Definition: Let $U = \{u_1, u_2, ..., u_n\}$ be the universe of discourse. Let π_A and π_B be two fuzzy data on U based on possibility distribution and $\pi_A (u_i)$, $u_i \in U$, denote the possibility that u_i is true. Let *Res* be a resemblance relation on domain U and α $(0 \le \alpha \le 1)$ be a threshold corresponding to *Res*. Then

$$SID_\alpha (\pi_A, \pi_B) = \sum_{i=1}^{n} \min_{u_i, u_j \in U \text{ and } Res_U (u_i, u_j) \ge \alpha} (\pi_B (u_i), \pi_A (u_j)) / \sum_{i=1}^{n} \pi_B (u_i)$$

and

$$SED_\alpha (\pi_A, \pi_B) = \min (SID_\alpha (\pi_A, \pi_B), SID_\alpha (\pi_B, \pi_A))$$

The notion of the semantic inclusion (or equivalence) degree of attribute values can be extended to the semantic equivalence degree of tuples. Let $t_i = <a_{i1}, a_{i2}, ..., a_{in}>$ and $t_j = <a_{j1}, a_{j2}, ..., a_{jn}>$ be two tuples in fuzzy relational instance r over schema R $(A_1, A_2, ..., A_n)$. The semantic inclusion degree of tuples t_i and t_j is denoted

$$SID_\alpha (t_i, t_j) = \min \{SID_\alpha (t_i [A_1], t_j [A_1]), SID_\alpha (t_i [A_2], t_j [A_2]), ..., SID_\alpha (t_i [A_n], t_j [A_n])\}.$$

The semantic equivalence degree of tuples t_i and t_j is denoted

$$SED_\alpha (t_i, t_j) = \min \{SED_\alpha (t_i [A_1], t_j [A_1]), SED_\alpha (t_i [A_2], t_j [A_2]), ..., SED_\alpha (t_i [A_n], t_j [A_n])\}.$$

Two types of fuzzy data redundancies: *inclusion redundancy* and *equivalence redundancy* can be classified and evaluated in fuzzy relational databases. Being different from the classical set theory, the condition $A = B$ is essentially the particular case of fuzzy data equivalence redundancy and the condition $A \supseteq B$ or $A \subseteq B$ is essentially the particular case of fuzzy data inclusion redundancy due to the data fuzziness. Here A and B are fuzzy sets. In general, the threshold should be considered when evaluating the semantic relationship between two fuzzy data.

Definition: Let π_A and π_B as well as α be the same as the above. Let β be a threshold. If SID_α $(\pi_A, \pi_B) \geq \beta$, it is said that π_B is *inclusively* redundant to π_A. If SED_α $(\pi_A, \pi_B) \geq \beta$, it is said that π_A and π_B are *equivalently* redundant to each other.

It is clear that equivalence redundancy of fuzzy data is a particular case of inclusion redundancy of fuzzy data. Considering the effect of resemblance relation in evaluation of semantic inclusion degree and equivalence degree, two fuzzy data π_A and π_B are considered equivalently α-β-redundant if and only if SED_α $(\pi_A, \pi_B) \geq \beta$. If SID_α $(\pi_A, \pi_B) \geq \beta$ and SID_α $(\pi_B, \pi_A) < \beta$, π_B is inclusively α-β-redundant to π_A.

When π_A and π_B are inclusively redundant or equivalently redundant, the removal of redundancy can be achieved by merging π_A and π_B and producing a new fuzzy data π_C. Following Zadeh's extension principle (Zadeh, 1975), the operation with an infix operator "θ" on π_A and π_B can be defined as follows.

$$\pi_A \ \theta \ \pi_B = \{\pi_A \ (u_i)/u_i \mid u_i \in U \wedge 1 \leq i \leq n\} \ \theta \ \{\pi_B \ (v_j)/v_j \mid v_j \in U \wedge 1 \leq j \leq n\}$$
$$= \{\max \ (\min \ (\pi_A \ (u_i), \ \pi_B \ (v_j))/ \ u_i \ \theta \ v_j)|u_i, v_j \in U \wedge 1 \leq i, j \leq n\}$$

Assume that π_A and π_B are α-β-redundant to each other, the elimination of duplicate could be achieved by merging π_A and π_B and producing a new fuzzy data π_C, where π_A, π_B and π_C are three fuzzy data on $U = \{u1, u2, \ldots, un\}$ based on possibility distribution and there is a resemblance relation Res_U on U. Then the following three merging operations are defined:

$$\pi_C = \pi_A \cup_f \pi_B = \{\pi_C \ (w)/w \mid (\exists \pi_A \ (ui)/ui) \ (\exists \pi_B \ (vj)/vj) \ (\pi_C \ (w) = \max \ (\pi_A$$
$(ui), \pi_B \ (vj)) \wedge (w = ui|_{\pi C \ (w) \ = \ \pi A \ (ui)} \vee w = vj|_{\pi C \ (w) \ = \ \pi B \ (vj)}) \wedge Res_U \ (ui, vj) \geq \alpha$
$\wedge ui, vj \in U \wedge 1 \leq i, j \leq n) \vee (\exists \pi_A \ (ui)/ui) \ (\forall \pi_B \ (vj)/vj) \ (\pi_C \ (w) = \pi_A \ (ui)$
$\wedge w = ui \wedge Res_U \ (ui, vj) \geq \alpha \wedge ui, vj \in U \wedge 1 \leq i, j \leq n) \vee (\exists \pi_B \ (vj)/(vj) \ (\forall$
$\pi_A \ (ui)/ui) \ (\pi_C \ (w) = \pi_B \ (vj) \wedge w = vj \wedge Res_U \ (ui, vj) \geq \alpha \wedge ui, vj \in U \wedge 1 \leq$
$$i, j \leq n)\},$$

$$\pi_C = \pi_A -_f \pi_B = \{\pi_C \ (w)/w \mid (\exists \pi_A \ (ui)/ui) \ (\exists \pi_B \ (vj)/vj) \ (\pi_C \ (w) = \max \ (\pi_A$$
$(ui) - \pi_B \ (vj), \ 0) \wedge w = ui \wedge Res_U \ (ui, vj) \geq \alpha \wedge ui, vj \in U \wedge 1 \leq i, j \leq n) \vee$
$(\exists \pi_A \ (ui)/ui) \ (\forall \pi_B \ (vj)/vj) \ (\pi_C \ (w) = \pi_A \ (ui) \wedge w = ui \wedge Res_U \ (ui, vj) < \alpha \wedge$
$$ui, vj \in U \wedge 1 \leq i, j \leq n)\}, \text{ and}$$

$$\pi_C = \pi_A \cap_f \pi_B = \{\pi_C (w)/w \mid (\exists \pi_A (ui)/ui) (\exists \pi_B (vj)/vj) (\pi_C (w) = \min (\pi_A$$
$$(ui), \pi_B (vj)) \wedge (w = ui|_{\pi_C (w) = \pi_A (ui)} \vee w = vj|_{\pi_C (w) = \pi_B (vj)}) \wedge Res_U (ui, vj) \geq \alpha$$
$$\wedge\ ui,\ vj \in U \wedge 1 \leq i, j \leq n)\}.$$

The processing of fuzzy value redundancy can be extended to that of fuzzy tuple redundancy. In a similar way, fuzzy tuple redundancy can be classified into *inclusion redundancy* and *equivalence redundancy* of tuples.

Definition: Let r be a fuzzy relation on the relational schema R (A_1, A_2, ..., A_n). Let $t = (\pi_{A1}, \pi_{A2}, ..., \pi_{An})$ and $t' = (\pi'_{A1}, \pi'_{A2}, ..., \pi'_{An})$ be two tuples in r. Let $\alpha \in [0, 1]$ and $\beta \in [0, 1]$ be two thresholds. The tuple t' is inclusively α-β-redundant to t if and only if min (SID$_\alpha$ (π_{Ai}, π'_{Ai})) $\geq \beta$ holds true ($1 \leq i \leq n$). The tuples t and t' are equivalently α-β-redundant if and only if min (SED$_\alpha$ (π_{Ai}, π'_{Ai})) $\geq \beta$ holds ($1 \leq i \leq n$).

7.3 The Fuzzy Nested Relational Database Model

The classical relational database model and its extension of imprecision and uncertainty do not satisfy the need of modeling complex objects with imprecision and uncertainty. So some researches have concentrated on introducing imprecise and uncertain information into NF^2 relational databases to deal with complex objects and uncertain data together. In (Levene, 1992; Roth *et al.*, 1989), a NF^2 database model with null values was presented. In (Yazici *et al*, 1999), uncertain null values, set values, range values (partial values and value intervals), and fuzzy values were all modeled in NF^2 data model. The extended algebra and the extended SQL-like query language were hereby defined. Also physical data representation of the model and the core operations that the model provides were introduced. Note that the fuzzy data in the extended NF^2 data model of (Yazici *et al*, 1999) are similarity-based (Buckles and Petry, 1982). In the following, we focus on the extended possibility-based representation of fuzzy data, where the fuzziness of data comes from possibility distributions and similarity (proximity or resemblance) relations over universes of discourse. We introduce extended possibility-based fuzzy data into nested relational databases.

An extended possibility-based fuzzy NF^2 relational schema is a set of attributes $R = (A_1, A_2, ..., A_n, pD)$ and their domains are $D_1, D_2, ..., D_n, D_0$, respectively, where D_i ($1 \leq i \leq n$) can be one of the following:

- The set of atomic values. For any an element $a_i \in D_i$, it is a typical simple crisp attribute value without imperfectness.

- The set of fuzzy subset. The corresponding attribute value is an extended possibility-based fuzzy data.
- The power set of the set in the first case. The corresponding attribute value, say a_i, is multivalued one with the form of $\{a_{i1}, a_{i2}, ..., a_{ik}\}$.
- The set of fuzzy subset. The corresponding attribute value, say a_i, is a fuzzy value represented by possibility distribution.
- The set of relation values. The corresponding attribute value, say a_i, is a tuple of the form $<a_{i1}, a_{i2}, ..., a_{im}>$ which is an element of $D_{i1} \times D_{i2} \times ... \times D_{im}$ ($m > 1$ and $1 \leq i \leq n$), where each D_{ij} ($1 \leq j \leq m$) may be a domain in all cases above and even the set of relation values.

The domain D_0 is a set of atomic values and each value is a crisp one from the range [0, 1], representing the possibility that the corresponding tuple is true in the NF^2 relation. We assume that the possibilities of all tuples are precisely one. Then for an attribute $A_i \in R$ ($1 \leq i \leq n$), its attribute domain is formally represented as follows:

$$\tau_i = dom \mid ndom \mid fdom \mid sdom \mid <B_1 : \tau_{i1}, B_2 : \tau_{i2}, ..., B_m : \tau_{im}>$$

where $B_1, B_2, ..., B_m$ are attributes.

A relational instance r over fuzzy NF^2 schema ($A_1 : \tau_1, A_2 : \tau_2, ..., A_n : \tau_n$) is a subset of Cartesian product $\tau_1 \times \tau_2 \times ... \times \tau_n$. A tuple in r with the form of $<a_1, a_2, ..., a_n>$ consists of n components. Each component a_i ($1 \leq i \leq n$) may be an atomic value, set value, fuzzy value, or another tuple.

Let us look at the preliminary design of a pressured air tank presented by Otto and Antonsson (1994). In the design, there are two design parameter values that need to be determined: length (l) and radius (r). In addition, there are four performance parameters in the design:

- Metal volume: the preference ranks of m are set because it is proportional to the cost;
- Tank capacity: the desired level of this performance ranks the preference;
- Overall height restriction: it is fuzzy;
- Overall radius restriction: it is also fuzzy.

Here, we focus on the modeling of these parameter values as well as its structure information in product data model using the fuzzy NF^2 database model. A possible database schema and an instance are partially represented in Table 7-1. Note that the attribute pD can be omitted from the fuzzy NF^2 relation when all tuples have value 1.0 on pD. It can be seen that $Tank\_Id$ and $Start\_data$ are crisp atomic-valued attributes, $Tank\_body$ is a relation-valued attribute, and $Responsibility$ is a set-valued attribute. In attribute $Tank\_body$, two component attributes $Volume$ and $Capacity$ are fuzzy ones. Here, "about 2.5e+03", "about 1.0e+06", "about 2.5e+04" and "about 1.0e+07" are all fuzzy values.

Table 7.1. Pressured air tank relation

Tank _ID	Tank_Body				Start_ Date	Respon- sibility
	Body_ID	Material	Volume	Capacity		
TA1	BO01	Alloy	about 2.5e+03	about 1.0e+06	01/12/99	John
TA2	BO02	Steel	about 2.5e+04	about 1.0e+07	28/03/00	{Tom, Mary}

As to the redundancies of tuples in a fuzzy NF^2 relation, first let us look at two values on a structured attribute $a_j = (A_{j1}: \pi_{Aj1}, A_{j2}: \pi_{Aj2}, ..., A_{jm}: \pi_{Ajm})$ and $a_j' = (A_{j1}: \pi_{Aj1}', A_{j2}: \pi_{Aj2}', ..., A_{jm}: \pi_{Ajm}')$, which consist of simple attribute values, crisp (atomic and set-valued) or fuzzy, on the schema $R (A_{j1}, A_{j2}, ..., A_{jm})$. There is a resemblance relation on each attribute domain D_{jk} $(1 \leq k \leq m)$ and $\alpha_{jk} \in [0, 1]$ $(1 \leq k \leq m)$ is the threshold on the resemblance relation. Let $\beta \in [0, 1]$ be a given threshold. a_j and a_j' are α-β-redundant if and only if for $k = 1, 2, ..., m$, min $(SE_{\alpha jk} (\pi_{Ajk}, \pi_{Ajk}')) \geq \beta$ holds true. min $(SE_{\alpha jk} (\pi_{Ajk}, \pi_{Ajk}'))$ $(1 \leq k \leq m)$ is called the equivalence degree of structured attribute values. Then the notion of equivalence degree of structured attribute values can be extended for the tuples in the fuzzy nested relations to assess tuple redundancies. Informally, any two tuples in a nested relation are redundant, if, for pair of the corresponding attribute values, the equivalence degree is greater than or equal to the threshold value. If the pair of the corresponding attribute values is simple, the equivalence degree is one for two values. For two values of structured attributes, however, the equivalence degree is one for structured attributes. Two redundant tuples t and t' are written $t \equiv t'$.

7.4 The Fuzzy Object-Oriented Databases

In order to represent and manipulate complex and uncertain data in databases, the fuzzy nested relational database model has been introduced (Yazici et al., 1999; Ma and Mili, 2002a). It should be noticed, however, that being the extension of relational data model, NF^2 data model is able to handle complex-valued attributes and may be better suited to some complex applications such as office automation systems, information retrieval systems and expert database systems (Yazici et al., 1999). But it is difficult for NF^2 data model to represent complex relationships among objects and attributes. Some advanced abstracts in data modeling (e.g., class hierarchy, inheritance, superclass/subclass, and encapsulation) are not supported by NF^2 data model, which are needed by many real applications. Therefore, in order to model fuzzy data and complex-valued attributes as well as complex relationships among objects, current efforts have being focused on object-oriented databases (*OODB*) with fuzzy information.

The incorporation of fuzzy information in object-oriented databases has received increasing attention, where fuzziness is witnessed at the levels of object instances and class hierarchies. Based on a similarity relationship, in (George et al., 1996), the range of attribute values is used to represent the set of allowed values for an attribute of a given class. Depending on the inclusion of the actual attribute values of the given object into the range of the attributes for the class, the membership degrees of an object to a class can be calculated. The weak and strong class hierarchies were defined based on monotone increase or decrease of the membership of a subclass in its superclass. Based on the extension of a graph-based object model, a fuzzy object-oriented data model was defined in (Bordogna et al., 1999). The notion of strength expressed by linguistic qualifiers was proposed, which can be associated with the instance relationship as well as an object with a class. Fuzzy classes and fuzzy class hierarchies were thus modeled in the object-oriented database system (OODB). In (Bordogna and Pasi, 2001), the definition of graph-based operations to select and browse a fuzzy object-oriented database that manages both crisp and fuzzy information was further proposed for the fuzzy graph-based model. A *UFO* (uncertainty and fuzziness in an object-oriented) database model was proposed in (Gyseghem and de Caluwe, 1998) to model fuzziness and uncertainty by means of fuzzy set theory and generalized fuzzy sets, respectively. The fact that the behavior and structure of the object are incompletely defined results in a gradual nature for the instantiation of an object. The partial inheritance, conditional inheritance, and multiple inheritances are permitted in fuzzy hierarchies. Based on possibility theory, vagueness and uncertainty were represented in class hierarchies in (Dubois et al., 1991), where the fuzzy ranges of the subclass attributes defined restrictions on that of the superclass attributes, and then the degree of inclusion of a subclass in the superclass was dependent on the inclusion between the fuzzy ranges of their attributes. Focusing on fuzzy types, Marín et al. (2001) discussed two different strategies for adding fuzzy types to an object-oriented database system and presented how the typical classes of an OODB can be used to represent a fuzzy type and the mechanisms of instantiation and inheritance can be modeled using this new type on an OODB. Recent efforts have been made to establish a consistent framework for a fuzzy object-oriented model based on the standard of the Object Data Management Group *(ODMG)* object data model (Cross *et al.*, 1997; Cross and Firat, 2000). In (de Tré and de Caluwe, 2003), an object-oriented database-modeling technique was presented based on the concept of a "level-2 fuzzy set" to deals with a uniform and advantageous representation of both perfect and imperfect real-world information. The same research also illustrates and discusses how the ODMG data model can be generalized to handle real-world data in a more advantageous way.

7.4.1 Fuzzy Objects and Fuzzy Classes

Objects model real-world entities or abstract concepts. Objects have properties that may be attributes of the object itself or relationships also known as associations between the object and one or more other objects. An object is fuzzy because of a lack of information. For example, an object representing a part in preliminary design for certain will also be made of *stainless steel*, *moulded steel*, or *alloy steel* (each of them may be connected with a possibility, say, 0.7, 0.5 and 0.9, respectively). Formally, objects that have at least one attribute whose value is a fuzzy set are fuzzy objects.

The notion of classes is introduced to represent a set of objects which have the same properties. In other words, such objects are gathered into a class. Generally classes are organized into hierarchies. Correspondingly we have two different viewpoints of a class (Dubois *et al.*, 1991):

- an extensional class, where the class is defined by the list of its object instances, and
- an intensional class, where the class is defined by a set of attributes and their admissible values.

In the object-oriented data modeling, a class can be also defined from its superclass by means of specialization mechanism or from its subclasses by means of generalization mechanism. So the subclass and superclass can be viewed as the special case of the intensional class above.

Based on the two viewpoints of classes, a class is fuzzy because of the following several reasons. First of all, some objects having similar properties are fuzzy ones. A class defined by these objects may be fuzzy. Meanwhile these objects belong to the class with membership degree of [0, 1]. Second of all, when a class is intensionally defined, the domain of an attribute may be fuzzy, which may result in a fuzzy class. Let us look at an example, class *Old equipment* is a fuzzy one because the domain of its attribute *Using period* is a set of fuzzy values such as *long*, *very long*, and *about 20 years*. Third of all, the subclass produced by a fuzzy class by means of specialization and the superclass produced by some classes (in which there is at least one class who is fuzzy) by means of generalization are also fuzzy.

The main difference between fuzzy classes and crisp classes is that the boundaries of fuzzy classes are imprecise. The imprecision in the class boundaries is caused by the imprecision of the values in the attribute domain. In the fuzzy *OODB*, classes are fuzzy because their attribute domains are fuzzy. The issue that an object fuzzily belongs to a class occurs since a class or an object is fuzzy. Similarly, a class is a subclass of another class with membership degree of [0, 1] because of the class fuzziness. In the *OODB*, the above-mentioned relationships are certain. The evaluations of fuzzy object-class relationships and fuzzy inheritance hierarchies are hereby the cores of information modeling in the fuzzy *OODB*. In the following discussion, we assume that the fuzzy attribute values of fuzzy objects and the fuzzy values in fuzzy attribute domains are represented by possibility distributions.

Now let us focus on the fuzzy object-class relationships. Since the objects and classes may be fuzzy, we can identify the following three kinds of fuzzy object-class relationships.

- Crisp class and fuzzy object. Although the class is precisely defined and has the precise boundary, an object is fuzzy since its attribute value(s) may be fuzzy. In this situation, the object may be related to the class with the special degree in [0, 1].
- Fuzzy class and crisp object. Being the same as the case in (b), the object may belong to the class with the membership degree in [0, 1].
- Fuzzy class and fuzzy object. In this situation, the object also belongs to the class with the membership degree in [0, 1].

Of course, we also have the object-class relationship of crisp class and crisp object. This situation is the same as the *OODB*, where the object belongs or not to the class certainly and can be seen the special case of fuzzy object-class relationships, where the membership degree of the object to the class is exactly one. It is not difficult to see that estimating the membership of an object to the class is crucial for fuzzy object-class relationship when classes are instantiated.

Based on the viewpoint of the intensional classes, in the classical *OODB*, determining if an object belongs to a class depends on if its attribute values are respectively included in the corresponding attribute domains of the class. Similarly, It is necessary to evaluate the degrees that the attribute domains of the class include the attribute values of the object in order to calculate the membership degree of an object to the class in a fuzzy object-class relationship. It should be noted, however, that in a fuzzy object-class relationship, only the inclusion degree of object values with respect to the class domains is not accurate for the evaluation of membership degree of an object to the class. It has been shown that the attributes play different role in the definition and identification of a class (Liu and Song, 2001). Some may be dominant and some not. So a weight w is assigned to each attribute of the class according to its importance by designer. Then the membership degree of an object to the class in a fuzzy object-class relationship can be calculated using the inclusion degree of object values with respect to the class domains, and the weight of attributes.

Formally, let C be a class on attribute set $\{A_1, A_2, ..., A_n\}$ with attribute domains $dom(A_1)$, $dom(A_2)$, ..., $dom(A_n)$ and o be an object also on attribute set $\{A_1, A2, ..., A_n\}$. Notation $o(A_i)$ denotes the attribute value of o on A_i ($1 \leq i \leq n$). Assume that the inclusion degree of $o(A_i)$ with respect to $dom(A_i)$ is denoted ID $(dom(A_i), o(A_i))$. As we know, $dom(A_i)$ is a set of crisp values in the classical *OODB* and may be a set of fuzzy subsets in the fuzzy databases. Therefore, in a uniform *OODB* for crisp and fuzzy information modeling, $dom(A_i)$ is the union of these two components, $dom(A_i) = cdom(A_i) \cup fdom(A_i)$, where $cdom(A_i)$ and $fdom(A_i)$ respectively denote the sets of crisp values and fuzzy subsets. Corresponding to the attribute domain, $o(A_i)$ may also be a crisp value or a fuzzy value. The following cases can be identified for evaluating ID $(dom(A_i), o(A_i))$.

(a) $o(A_i)$ is a fuzzy value. Let $fdom(A_i) = \{f_1, f_2, ..., f_m\}$, where f_j ($1 \leq j \leq m$) is a fuzzy value, and $cdom(A_i) = \{c_1, c_2, ..., c_k\}$, where c_l ($1 \leq l \leq k$) is a crisp value. Then

$$\text{ID}(dom(A_i), o(A_i)) = \max(\text{ID}(cdom(A_i), o(A_i)), \text{ID}(fdom(A_i), o(A_i))) = \max(\text{SID}(\{1.0/c_1, 1.0/c_2, ..., 1.0/c_k\}, o(A_i)), \max_j(\text{SID}(f_j, o(A_i)))),$$

where SID (x, y), as shown in Section 7.2, is used to calculate the degree that fuzzy value x include fuzzy value y.

(b) $o(A_i)$ is a crisp value. Then

$$\text{ID}(dom(A_i), o(A_i)) = 1 \text{ if } o(A_i) \in cdom(A_i) \text{ else ID}(dom(A_i), o(A_i)) = \text{ID}(fdom(A_i), \{1.0/o(A_i)\}).$$

Now, we define the formula to calculate the membership degree of the object o to the class C as follows, where $w(A_i(C))$ denotes the weight of attribute A_i to class C. Then we have

$$\mu_C(o) = \frac{\sum_{i=1}^{n} ID\left(dom\left(A_i\right), o\left(A_i\right)\right) \times w\left(A_i\left(C\right)\right)}{\sum_{i=1}^{n} w\left(A_i\left(C\right)\right)}$$

It should be noticed, however, that in the above-given determination that an object belongs to a class fuzzily, it is assumed that the object and the class have the same attributes, namely, class C is with attributes $\{A_1, A_2, \ldots, A_n\}$ and object o is on $\{A_1, A_2, \ldots, A_n\}$ also. Such an object-class relationship was called direct object-class relationship in (Ma, 2005) and furthermore one kind of special object-class relationship, called indirect object-class relationship, was proposed. In the indirect object-class relationship, the object and the class have different attribute sets. Therefore, the above-given determination method for the direct object-class relationship can not be applied for the indirect object-class relationship. In the following, we present how to calculate the membership degree of an object to the class in an indirect object-class relationship.

Let C be a class with attributes $\{A_1, A_2, \ldots, A_k, A_{k+1}, \ldots, A_m, A_{m+1}, \ldots, A_n\}$ and o be an object on attributes $\{A_1, A_2, \ldots, A_k, A'_{k+1}, \ldots, A'_m, B_{m+1}, \ldots, B_p\}$. Here attributes $A'_{k+1}, \ldots,$ and A'_m are overridden from $A_{k+1}, \ldots,$ and A_m, or $A_{k+1}, \ldots,$ and A_m are overridden from $A'_{k+1}, \ldots,$ and A'_m. Attributes $A_{m+1}, \ldots,$ and A_n and $B_{m+1}, \ldots, B_p$ are special in $\{A_1, A_2, \ldots, A_k, A_{k+1}, \ldots, A_m, A_{m+1}, \ldots, A_n\}$ and $\{A_1, A_2, \ldots, A_k, A'_{k+1}, \ldots, A'_m, B_{m+1}, \ldots, B_p\}$, respectively. Then we have

$$\mu_C(o) = \frac{\sum_{i=1}^{k} ID\left(dom\left(A_i\right), o\left(A_i\right)\right) \times w\left(A_i\left(C\right)\right) + \sum_{j=k+1}^{m} ID\left(dom\left(A_j\right), o\left(A_j'\right)\right) \times w\left(A_j\left(C\right)\right)}{\sum_{i=1}^{n} w\left(A_i\left(C\right)\right)}$$

Since an object may belong to a class with membership degree in [0, 1] in fuzzy object-class relationship, it is possible that an object belongs to different classes with different membership degrees simultaneously in fuzzy object-class relationships. The situation, called multiple membership of object in (Ma, 2005), only arises in the fuzzy *OODB*.

7.4.2 Fuzzy Inheritance Hierarchies

In the classical *OODB*, subclasses are produced from their superclass by means of inheriting some attributes and methods of the superclass, overriding some attributes and methods of the superclass, and defining some new attributes and methods. Therefore, a subclass is the specialization of its superclass and any one object belonging to the subclass must belong to the superclass. This characteristic can be used to determine if two classes have subclass/superclass relationship in the classical *OODB*.

In the fuzzy *OODB*, a subclass produced from a fuzzy class, called fuzzy superclass, may be fuzzy. Then the subclass/superclass relationship is fuzzy. That means that a class is a subclass of another class with membership degree of [0, 1]. Correspondingly, the method used in the classical *OODB* for determining the subclass/superclass relationship should be modified as follows.

- for any (fuzzy) object, if the member degree that it belongs to the subclass is less than or equal to the member degree that it belongs to the superclass, and
- the member degree that it belongs to the subclass is great than or equal to the given threshold.

Formally, let C_1 and C_2 be (fuzzy) classes, o be an object, and β be a given threshold. Then C_2 is a subclass of C_1 if

$$(\forall o)\ (\beta \leq \mu_{C2}\ (o) \leq \mu_{C1}\ (o))$$

Here $\mu_{C1}\ (o)$ and $\mu_{C2}\ (o)$ are the member degrees that o belongs to C_1 and C_2, respectively. The membership degree that C_2 is a subclass of C_1 should be $\min_{\mu_{C2}\ (o) \geq \beta}\ (\mu_{C2}\ (o))$.

In the above, based on the inclusion degree of objects to the class, we give a method for assessing the fuzzy subclass/superclass relationships in the fuzzy *OODB*. It is clear that this method can not be used if there is no any object available. We should develop other method to determine the relationships between fuzzy subclass and superclass. We can calculate the inclusion degree of a (fuzzy) subclass with respect to the (fuzzy) superclass according to the inclusion degree of the attribute domains of the subclass with respect to the attribute domains of the superclass as well as the weight of attributes. In the following, we give the method for evaluating the inclusion degree of fuzzy attribute domains.

Let C_1 and C_2 be (fuzzy) classes with attribute set $\{A_1, A_2, ..., A_k, A_{k+1}, ..., A_m\}$ and $\{A_1, A_2, ..., A_k, A'_{k+1}, ..., A'_m, A_{m+1}, ..., A_n\}$, respectively. It is clear that attributes $A_1, A_2, ...,$ and A_k in C_2 are directly inherited from $A_1, A_2, ...,$ and A_k in C_1, attributes $A'_{k+1}, ...,$ and A'_m in C_2 are overridden from $A_{k+1}, ...,$ and A_m in C_1, and attributes $A_{m+1}, ...,$ and A_n in C_2 are special. For each attribute in C_1 or C_2, say A_i, there is a domain, denoted $dom\ (A_i)$. As shown above, $dom\ (A_i)$ should be $dom\ (A_i) = cdom\ (A_i) \cup fdom\ (A_i)$, where $cdom\ (A_i)$ and $fdom\ (A_i)$ denote the sets of crisp values and fuzzy subsets, respectively. Let A_i and A_j be attributes of C_1 and C_2, respectively. The inclusion degree of $dom\ (A_j)$ with respect to $dom\ (A_i)$ is denoted by ID (dom (A$_i$), dom (A$_j$)). We identify the following cases and investigate the evaluation of ID (dom (A$_i$), dom (A$_j$)):

- when $i \neq j$ and $1 \leq i, j \leq k$, ID (dom (A$_i$), dom (A$_j$)) = 0,
- when $i = j$ and $1 \leq i, j \leq k$, ID (dom (A$_i$), dom (A$_j$)) = 1, and
- when $i = j$ and $k + 1 \leq i, j \leq m$, ID (dom (A$_i$), dom (A$_j$)) = ID (dom (A$_i$), dom (A'$_i$)) = max (ID (dom (A$_i$), $cdom$ (A'$_i$)), ID (dom (A$_i$), $fdom$ (A'$_i$))).

Now let us focus on ID (dom (A$_i$), $cdom$ (A'$_i$)) and ID (dom (A$_i$), $fdom$ (A'$_i$)). Let $fdom$ (A'$_i$) = {$f_1, f_2, \dots, f_m$}, where f_j ($1 \leq j \leq m$) is a fuzzy value, and $cdom$ (A'$_i$) = {$c_1, c_2, \dots, c_k$}, where c_l ($1 \leq l \leq k$) is a crisp value. We can consider {$c_1, c_2, \dots, c_k$} as a special fuzzy set {$1.0/c_1, 1.0/c_2, \dots, 1.0/c_k$}. Then we have

$$\text{ID} (dom (A_i), cdom (A'_i)) = \text{ID} (dom (A_i), \{1.0/c_1, 1.0/c_2, \dots, 1.0/c_k\}) \text{ and}$$
$$\text{ID} (dom (A_i), fdom (A'_i)) = \max_j (\text{ID} (dom (A_i), f_j)).$$

Based on the inclusion degree of attribute domains of the subclass with respect to the attribute domains of its superclass as well as the weight of attributes, we can define the formula to calculate the degree to which a fuzzy class is a subclass of another fuzzy class. Let C_1 and C_2 be the same as the above and w (A$_i$) denote the weight of attribute A$_i$. Then the degree that C_2 is the subclass of C_1, denoted μ (C_1, C_2) is defined as follows.

$$\mu (C1, C2) = \frac{\sum_{i=1}^{m} ID (dom (A_i (C1)), dom (A_i (C2))) \times w (A_i)}{\sum_{i=1}^{m} w (A_i)}$$

Now let us look at the multiple inheritance of class. In the multiple inheritance of class, ambiguity arises when more than one of the superclasses have common attributes and the subclass does not declare explicitly the class from which the attribute is inherited. Formally, let class C be a subclass of classes C_1 and C_2. Assume that the attribute A$_i$ in C_1, denoted by A$_i$ (C_1), is common to the attribute A$_i$ in C_2, denoted A$_i$ (C_2). Then

- If dom (A$_i$ (C_1)) and dom (A$_i$ (C_2)) are identical, there does not exist a conflict in the multiple inheritance hierarchy and C inherits attribute A$_i$ directly.
- If dom (A$_i$ (C_1)) and dom (A$_i$ (C_2)) are not identical, however, the conflict occurs.

For the second situation, which one in A$_i$ (C_1)) and A$_i$ (C_2) is inherited by C dependents on the following rule:

If ID (dom (A$_i$ (C$_1$)), dom (A$_i$ (C$_2$))) $\times w$ (A$_i$ (C$_1$)) > ID (dom (A$_i$ (C$_2$)), *dom (A_i (C_1))) × w (A_i (C_2)), then A_i (C_1) is inherited by C, else A_i (C_2) is inherited by C.*

It should be noted that not being the same as the situation in classical object-oriented database systems, the subclass has different degrees with respect to different superclasses in fuzzy multiple inheritance hierarchy.

7.4.3 The Fuzzy Object-Oriented Database Model

The specification of a class includes the definition of *ISA* relationships, attributes and method implementations in the classical *OODB*. In the fuzzy *OODB*, however, the classes may be fuzzy. Accordingly, in the fuzzy *OODB*, an object belongs to a class with a membership degree of [0, 1] and a class is the subclass of another class with degree of [0, 1] also. To the purpose of specifying a fuzzy class, some additional definitions are needed in addition to the definitions given in the classical class. First, the weights of attributes to the class must be given. Also a new attribute should be added into the class to indicate the membership degree to which an object belongs to the class. If the class is a fuzzy subclass, its superclasses and the degree that the class is the subclass of the superclass should be illustrated in the specification of the class. Finally, in the definition of a fuzzy class, fuzzy attributes should be explicitly indicated.

Formally, the definition of a fuzzy class is shown as follows.

```
CLASS class name WITH DEGREE OF degree
    INHERITS superclass_1 name WITH DEGREE OF degree_1
    ...
    INHERITS superclass_k name WITH DEGREE OF degree_k
    ATTRIBUTES
        Attribute_1 name: [FUZZY] DOMAIN dom_1: TYPE OF type_1 WITH
DEGREE OF degree_1
        ...
        Attribute_m name: [FUZZY] DOMAIN dom_m: TYPE OF type_m WITH
DEGREE OF degree_m
        Membership_Attribute name: membership_degree
    WEIGHT
        w (Attribute_1 name) = w_1
        ...
        w (Attribute_m name) = w_m
    METHODS
        ...
END
```

In the classical *OODB*, the data types of attributes include simple types (e.g., integer, real, Boolean, and string) and complex types (e.g., set type and object type). The fuzzy *OODB* may contain fuzzy attributes which are explicitly indicated in a class definition. The data types of the fuzzy attributes are fuzzy type based simple types or complex types, which allows the representation of descriptive form of imprecise information. In the class definition above, we declare only the base type (e.g., integer) of fuzzy attributes and the fuzzy domain. A fuzzy domain is a set of possibility distributions or fuzzy linguistic terms (each fuzzy term is associated with a membership function). For example, a fuzzy attribute *Age* may be declared as follows.

Age: FUZZY DOMAIN {*very young, young, old, very old*}: TYPE OF *integer*

Then an object attribute with fuzzy type will have either a crisp value or a fuzzy value given in the type definition. For example, *Age* = 21 or *Age* = *young*.

7.5 Summary

In this chapter, several major fuzzy database models have been presented. These fuzzy database models include the fuzzy relational databases, the fuzzy nested relational databases and the fuzzy object-oriented databases. In particular, a method of quantitative measure of the semantic relationships of fuzzy data has been developed. The quantitative measure of fuzzy data can overcome counterintuitive results when existing methods are applied.

The semantic measure method of fuzzy data developed in this chapter can be used not only for the definitions of the data dependencies, data redundancy, and relational operations in the fuzzy relational databases (Ma *et al.*, 2002; Ma and Mili, 2002b), but also in the fuzzy nested relational database model. In addition, the semantic measure method of fuzzy data developed in this chapter has been applied for the calculations of the object-class relationship and inheritance hierarchy in the fuzzy object-oriented databases. On the basis, a generic model for the fuzzy object-oriented databases has been developed in the chapter.

References

Bordogna, G. and Pasi, G. (2001), Graph-based interaction in a fuzzy object oriented database, International Journal of Intelligent Systems, 16 (7): 821-841.

Bordogna, G., Pasi, G. and Lucarella, D. (1999), A fuzzy object-oriented data model for managing vague and uncertain information, International Journal of Intelligent Systems, 14: 623-651.

Bosc, P. and Pivert, O. (1997), On the comparison of imprecise values in fuzzy databases, Proceedings of the 1997 IEEE International Conference on Fuzzy Systems, 2: 707-712.

Bosc, P. and Pivert, O. (2003), On the impact of regular functional dependencies when moving to a possibilistic database framework, Fuzzy Sets and Systems, 140 (1): 207-227.

Bosc, P. and Pivert, O. (1995), SQLf: a relational database language for fuzzy querying, IEEE Transactions on Fuzzy Systems, 3 (1): 1-17.

Bosc, P., Dubois, D. and Prade, H. (1998), Fuzzy functional dependencies and re-dundancy elimination, Journal of the American Society for Information Science, 49 (3), 217-235.

Buckles, B. P. and Petry, F. E. (1982), A fuzzy representation of data for relational database, Fuzzy Sets and Systems, 7 (3): 213-226.

Chen, G. Q. (1999), Fuzzy Logic in Data Modeling; Semantics, Constraints, and Database Design, Kluwer Academic Publisher.

Chen, G. Q., Vandenbulcke, J. and Kerre, E. E. (1992), A general treatment of data redundancy in a fuzzy relational data model, Journal American Society of Information Sciences, 43 (3): 304-311.

Cross, V. and Firat, A. (2000), Fuzzy objects for geographical information systems, Fuzzy Sets and Systems, 113: 19-36.

Cross, V., de Caluwe, R. and Vangyseghem, N. (1997), A perspective from the fuzzy object data management group (FODMG), Proceedings of the 1997 IEEE International Conference on Fuzzy Systems, 2: 721-728.

Cubero, J. C. and Vila, M. A. (1994), A new definition of fuzzy functional dependency in fuzzy relational databases, International Journal of Intelligent Systems, 9 (5): 441-448.

de Tré, G. and de Caluwe, R. (2003), Level-2 fuzzy sets and their usefulness in object-oriented database modelling, Fuzzy Sets and Systems, 140 (1): 29-49.

Dubois, D., Prade, H. and Rossazza, J. P. (1991), Vagueness, typicality, and un-certainty in class hierarchies, International Journal of Intelligent Systems, 6: 167-183.

George, R., Srikanth, R., Petry, F. E. and Buckles, B. P. (1996), Uncertainty man-agement issues in the object-oriented data model, IEEE Transactions on Fuzzy Systems, 4 (2): 179-192.

Gyseghem, N. V. and de Caluwe, R. (1998), Imprecision and incertainty in UFO database model, Journal of the American Society for Information Science, 49 (3): 236-252.

Levene, M., 1992, The Nested Universal Relation Database Model, Lecture Notes in Computer Science No. 595, Springer-Verlag, Berlin.

Liu, W. Y. (1997), Fuzzy data dependencies and implication of fuzzy data de-pendencies, Fuzzy Sets and Systems, 92 (3): 341–348.

Liu, W. Y. and Song, N. (2001), The fuzzy association degree in semantic data models, Fuzzy Sets and Systems, 117 (2): 203-208.

Ma, Z. M. (2005), Fuzzy Database Modeling with XML, Springer.

Ma, Z. M., Zhang, W. J., Ma, W. Y. and Mili, F. (2002), Data dependencies in extended possibility-based fuzzy relational databases, International Journal of Intelligent Systems, 17 (3): 321-332.

Ma, Z. M. and Mili, F. (2002a), An extended possibility-based fuzzy nested relational database model and algebra, IFIP International Federation for Information Processing (KLUWER Academic Publishers), 221: 285-288.

Ma, Z. M. and Mili, F. (2002b), Handling fuzzy information in extended possibility-based fuzzy relational databases, International Journal of Intelligent Systems, 17 (10): 925-942.

Ma, Z. M., Zhang, W. J. and Ma, W. Y. (1999), Assessment of data redundancy in fuzzy relational databases based on semantic inclusion degree, Information Processing Letters, 72 (1-2): 25-29.

Ma, Z. M., Zhang, W. J. and Ma, W. Y. (2000), Semantic measure of fuzzy data in extended possibility-based fuzzy relational databases, International Journal of Intelligent Systems, 15 (8): 705-716.

Marín, N., Pons, O. and Vila, M. A. (2001), A strategy for adding fuzzy types to an object-oriented database system, International Journal of Intelligent Systems, 16 (7): 863-880.

Otto, K. N. and Antonsoon, E. K. (1994), Design parameter selection in the presence of noise, Research in Engineering Design, 6 (4): 234-246.

Petry, F. E. (1996), Fuzzy Databases: Principles and Applications, Kluwer Academic Publisher.

Prade, H. and Testemale, C. (1984), Generalizing database relational algebra for the treatment of incomplete or uncertain information and vague queries, Information Sciences, 34: 115-143.

Raju, K. V. S. V. N. and Majumdar, A. K. (1988), Fuzzy functional dependencies and lossless join decomposition of fuzzy relational database system, ACM Transactions on Database Systems, 13(2): 129-166.

Roth, M. A., Korth, H. F. and Silberschatz, A., 1989, Null values in nested relational databases, *Acta Informatica*, 26: 615-642.

Rundensteiner, E. A., Hawkes, L. W. and Bandler, W. (1989), On nearness measures in fuzzy relational data models, International Journal of Approximate Reasoning, 3: 267-98.

Shenoi, S. and Melton, A. (1989), Proximity relations in the fuzzy relational databases, Fuzzy Sets and Systems, 31 (3): 285-296, 1989.

Sözat, M. I. and Yazici, A. (2001), A complete axiomatization for fuzzy functional and multivalued dependencies in fuzzy database relations, Fuzzy Sets and Systems, 117 (2): 161-181.

Takahashi, Y. (1993), Fuzzy database query languages and their relational completeness theorem, IEEE Transactions on Knowledge and Data Engineering, 5 (1): 122-125.

Umano, M. and Fukami, S. (1994), Fuzzy relational algebra for possibility-distribution-fuzzy-relational model of fuzzy data, Journal of Intelligent Information Systems, 3: 7-27.

Yazici, A. and George, R. (1999), Fuzzy Database Modeling, Physica-Verlag.

Yazici, A., Buckles, B. P. and Petry, F. E. (1992), A survey of conceptual and logical data models for uncertainty management, Fuzzy Logic for Management of Uncertainty, John Wiley and Sons Inc., 607-644.

Yazici, A., Soysal, A., Buckles, B. P. and Petry, F. E., 1999, Uncertainty in a nested relational database model, *Data & Knowledge Engineering*, 30: 275–301.

Zadeh, L. A. (1965), Fuzzy sets, Information and Control, 8 (3): 338-353.

Zadeh, L. A. (1975), The concept of a linguistic variable and its application to approximate reasoning, Information Sciences, 8: 119-249 & 301-357; 9: 43-80.

Zadeh, L. A. (1978), Fuzzy sets as a basis for a theory of possibility, Fuzzy Sets and Systems, 1 (1): 3-28.

Zemankova M. and Kandel A. (1985), Implementing imprecision in information systems, Information Sciences, 37 (1-3): 107-141.

8 Conceptual Designs of the Fuzzy Databases

8.1 Introduction

Database modeling in databases can be carried out at two different levels: *conceptual data modeling* and *logical database modeling*. Therefore, we have conceptual data models and logical database models for database modeling, respectively. Conceptual data models can capture and represent rich and complex semantics at a high abstract level (Shoval and Frumermann, 1994; Halpin, 2002). Database modeling generally starts from the conceptual data models and then the developed conceptual data models are mapped into the logical database models. This conversion is called *conceptual design of databases*. Various conceptual data models have been used for conceptual design of databases. For example, relational databases were designed by first developing a high-level conceptual data model, ER model, and then the developed conceptual model is mapped to an actual implementation of relational databases (Teorey *et al.*, 1986). In addition to ER model, the IFO model developed in (Abiteboul and Hull, 1987) was extended into a formal object model IFO_2 and then the IFO_2 model was mapped into object-oriented databases in (Poncelet *et al.*, 1993).

It has been shown that there have been several proposals for extending relational database model as well as nested relational database model in order to represent and query fuzzy data. Also current efforts have been concentrated on fuzzy object-oriented databases and some related notions. Compared with the studies of the fuzzy database models, little work has been done in modeling fuzziness in conceptual data models. It is particularly true in developing design methodologies for implementing fuzzy databases. This chapter will investigate the conceptual designs of the fuzzy databases, including the conceptual designs of the relational databases using the fuzzy ER model and the conceptual designs of the fuzzy object-oriented databases using the fuzzy EER model.

Z. Ma: *Fuzzy Database Modeling of Imprecise and Uncertain Engineering Information*, StudFuzz **195**, 159–166 (2006)
www.springerlink.com

8.2 Mapping the Fuzzy ER Model to the Fuzzy Relational Database Model

Following the extension of the ER model in (Zvieli and Chen 1986), Chaudhry *et al.* (1999) proposed a method for design the fuzzy relational databases (FRDBs) to convert crisp database into fuzzy ones.

8.2.1 Mapping the Fuzzy ER Model to a Relational Implementation

In (Chaudhry *et al.*, 1999), it was assumed that the fuzzy ER model is mapped to a conventional crisp database that only allows simple data types. To this purpose, the entities and relationships with fuzzy attributes should be mapped to conventional crisp databases, which allow only atomic attributes. Identifying various entities and relationships, several kinds of mappings/transformations were proposed in (Chaudhry *et al.*, 1999).

Mappings of fuzzy entities and fuzzy many-to-many relationships. The fuzzy entities and fuzzy many-to-many relationships without any fuzzy attributes can be mapped into relational instances, i.e., relational tables. But these relations must contain one additional attribute, namely, membership degree attribute, which is used to indicate, for each special entity/relationship instance, the membership degree to which the instance belongs to the entity/relationship. It is clear that the mappings above can be used only for the entities and relationships with the second level of fuzziness in the fuzzy ER model of (Zvieli and Chen 1986).

Mappings of fuzzy one-to-one and one-to-many relationships. Generally speaking, a relationship of ER model is mapped into a separate relational table. But in traditional ER model, the implementation of a one-to-one or one-to-many relationship with no attributes does not need a separate relational table as the information about cardinality in the relationships can be kept in one of the entity tables by storing a foreign. In the fuzzy ER model, however, a fuzzy relationship necessarily has at least one attribute, which models the fuzziness in the relation, namely, the membership degree. Therefore, a fuzzy one-to-one or one-to-many relationship should be mapped to a separate relational table rather than to the entity table. Also the mappings above can be used only for the one-to-one and one-to-many relationships with the second level of fuzziness in the fuzzy ER model of (Zvieli and Chen 1986).

Mappings of entities with fuzzy attributes. The entities with fuzzy attributes can be mapped into separate relational tables. The fuzzy attributes will be mapped according to the technique described blow. Here the fuzzy attributes mean the third level of fuzziness in the fuzzy ER model of (Zvieli and Chen 1986).

Normalizing entities and relationships with fuzzy attributes. As we know, the domain of a fuzzy attribute is a set of possibility distributions (or equivalently fuzzy sets). It is clear that the domain is no-atomic. In traditional relational databases, it is required that all attributes have atomic domains, namely, all elements of each domain are considered to be indivisible units. Therefore, basically we have two kinds of fuzzy relational databases: the first one is the database which allows possibility distributions as data type and the second one is the database which does not allow possibility distributions as data type. In the first kind of fuzzy relational databases, the fuzzy attributes in the fuzzy ER model can be directly mapped to the attributes of the entity relational table mapped from the corresponding entity. But the second kind of fuzzy relation will be not in first normal form (1NF). The entities with fuzzy attributes must be normalized if the databases do not allowed possibility distributions as a data type.

In (Chaudhry *et al.*, 1999), a strategy was given to normalize a possibilistic entity. Let R be a relation with attributes $(K, A_1, ..., A_l)$, with non-fuzzy key K, non-key (fuzzy or non-fuzzy) attributes A_m for $m = 1, ..., l$. For each A_m ($1 \leq m \leq l$), which is fuzzy (i.e., its domain is a fuzzy set), a new relation $A_m = (K_m, value, \mu)$ is created. Then R is modified to R' by replacing A_m in R with K_m in R'. Here *value* and μ respectively model the elements and associated degrees of the fuzzy attribute A_m, and K_m is an atomic identifier in R' for A_m. Formally

$$r = \ <k, a_1, ..., a_m, ..., a_l> \ \in R \text{ with } a_m = \{\mu(a_{m1})/a_{m1}, ..., \mu(a_{mn})/a_{mn}\},$$

add the collection of tuples $a^F_{mp} = \ <k_m, a_{mp}, \mu(a_{mp})>$ for $p = 1, ..., n$ to the new relation A_m and add the instance $<k, a_1, ..., k_m, ..., a_l>$ to $R\bullet$.

As indicated by Chaudhry *et al.* (1999), this method can be used only for attributes whose domain consists of fuzzy sets with finite cardinality. We cannot use this method to normalize relations with attributes whose domain includes fuzzy sets with infinite cardinality because this will require an infinite number of tuples.

8.2.2 Fuzzy ER Conceptual Design Methodology

Originally the ER design methodology was proposed in (Teorey *et al.*, 1986). Incorporating the mapping of the fuzzy ER model to relational databases, Chaudhry *et al.* (1999) gave FERM, a design methodology for mapping a fuzzy ER data model to a crisp relational database in four steps, which are

- *Step* 1: constructing a fuzzy ER data model
- *Step* 2: transforming the ER model constructs to relational tables
- *Step* 3: normalization of the relations
- *Step* 4: ensuring correct interpretation of the fuzzy relational operators

The construction of a fuzzy ER data model starts from the construction of the traditional ER model. Then the three levels of fuzziness in (Zvieli and Chen 1986) are attached to the entities, relationships and attributes that are fuzzy. The transformation of the ER model to relational tables first identifies crisp entities and crisp relationships and they can be mapped to relational tables as described in (Teorey *et al.*, 1986). Then the fuzzy entities and fuzzy relationships with the second level of fuzziness in (Zvieli and Chen 1986) are transformed to individual relational tables with one attribute, i.e., the membership degree attribute. Finally the entities with the third level of fuzziness in (Zvieli and Chen 1986) are transformed to individual relational tables. It should be noticed, however, that the mapped relations with the fuzzy attribute values should be normalized if the databases do not allow possibility distributions as a data type. The normalization also includes normalizing the mapped relations by using some integrity constraints such as functional and multi-valued dependencies. The step of ensuring correct interpretation of the fuzzy relational operators is related to the operations on the data rather than the data itself. This step is needed because there is no commercially available fuzzy DBMS at current stage. The fuzzy database designer must make sure that the results obtained from a traditional RDBMS conform to the semantics implied by the various fuzzy relational operators.

8.3 Mapping the Fuzzy EER Model to the Fuzzy Object-Oriented Database Model

After constructing the fuzzy extended entity-relationship (FEER) model, we can map it to the fuzzy object-oriented database (FOODB) model. This mapping follows the similar principle as the proposed approach from the EER model to the OODB schema (Fong, 1995). Note that the ways of translation are different. In the following, we give the formal approach for the mapping of the fuzzy EER model to the fuzzy object-oriented database model.

8.3.1 Transformation of Entities

Generally speaking, the entities in the FEER model are transformed to the classes in the FOODB and the attributes of entities may be considered as the attributes of the corresponding class. To this purpose, we distinguish four kinds of entities in the FEER model as follows.

• Entities without any fuzziness at the three levels.
• Entities with the fuzziness only at the third level.
• Entities with the fuzziness at the second level.
• Entities with the fuzziness at the first level.

The first two kinds of entities can be directly mapped to the classes of the FEER. But for the third kinds of entities, an additional attribute should be added into each class transformed from the corresponding entity, which is used to denote the possibility that the class instances belong to the class.

We should pay extreme attention to the transformation of the entities with membership degree and the entities which attributes have membership degree. In addition to the transformation from the entities to the classes and from the attributes to the attributes, for such an entity or attribute, an additional constraint rule is needed to add into the class mapped from the entity, which is used to indicate the possibility that the class belongs to the corresponding database model, or the attribute of the class belong to the corresponding class. Also it is possible that there exist two kinds of fuzziness at the first and second levels in an entity simultaneously.

It should be pointed out, however, that the entities discussed above are not subclasses. The mapping of the entities of subclass will be discussed below. In addition, a weak entity in the FEER model depends on its owner entity, which can be mapped to a class according to the methods mentioned above such that there is existence dependency between two classes. The own statements should be included in the method parts of two classes, respectively.

8.3.2 Transformation of Relationships

Generally speaking, a relationship in the EER model can be mapped into an association in the fuzzy OO schema. The association describes a group of pointer as an attribute of object that combines an explicit reference to another object. Considering the constraint of cardinality, such attributes in two associated objects can be single-valued or multivalued ones.

Four kinds of relationships can be found in the FEER model.

• Relationships without any fuzziness at the three levels

- Relationships with the fuzziness only at the third level.
- Relationships with the second levels of fuzziness.
- Relationships with the first level of fuzziness.

A relationship is connected with two entities. For the first three kinds of relationships, two additional attributes can directly be appended into the classes transformed from the corresponding two entities, respectively. Such additional attributes are used to indicate the association of objects in the classes. For the transformation of the relationships with membership degree, namely, the fourth relationships above, however, the constraint rules indicating the possibility that the appended attributes belong to the corresponding classes are needed in addition to the transformation from relationships to association. Note that for a relationship of ownership, the same transformation can be conducted because the own statements have been included in the method parts of two classes, respectively.

8.3.3 Transformation of Generalizations

For the generalization, the variances among entities are suppressed and their commonalties are identified by generalizing them into one single class. The created class is called superclass and original entities with each of its unique differences are special subclasses. If all subclasses are crisp, the superclass must be crisp. Then the transformation from the EER model to the OODB is simply to map each entity to a class and indicate the inheritances among superclass and subclasses. Now we focus on the transformation of generalization with fuzzy entities.

Let us look at the entities with the second level of fuzziness. Assume that several fuzzy entities with the second level of fuzziness will be generalized into a superclass. Then the generalized superclass may be a crisp entity or a fuzzy entity with the second level of fuzziness. *"Young person"*, *"Mid-aged person"*, and *"Old person"*, for example, are three fuzzy entities with the second level of fuzziness. These three fuzzy entities can be generalized into a crisp entity *"Person"*. As to the transformation of the entities in the subclass-superclass relationship to the OO schema, the entity as the superclass can be converted to a class according to the approaches given above and each entity as a subclass can also be mapped to a class, which must be indicated to be a subclass and inherits the properties and operations of the superclass.

8.3.4 Transformation of Specializations

For the specialization, an entity or a fuzzy entity (superclass) can be specialized into a fuzzy entity (subclass). An entity *Car*, for example, is specialized into two fuzzy entities *"New Car"* and *"Old Car"*. Based on the same methods as the transformation of the generalization above, the entity as the superclass is converted to a class and the entity as a subclass is mapped to a class, which is indicated to be a subclass and inherits the properties and operations of the superclass. If a subclass have multiple superclasses, being as the same as classical OODB, the subclass should inherit the properties and operations of all superclasses and then the conflicts in multiple inheritance are resolved according to some rules.

8.3.5 Transformation of Categorizations

The categorization is concerned with the issue of selective inheritance. Essentially, the categorization shows the uncertainty that which entity in the categorization will take place in the schema is unknown currently. The entities, fuzzy or not, in the categorization can respectively be mapped into classes following the methods given above. The categorization entity also follows the transformation. Some additional attributes, however, should be added into the corresponding class. These attributes are a set of all attributes of the entities in the categorization. Of course, an instance of the categorization class must have null values on some attributes.

8.3.6 Transformation of Aggregations

Each aggregation in the FEER model can be mapped into an aggregation (or composite object) of the FOODB schema with component classes, which may be a crisp class or a fuzzy class, depending on if its component classes are fuzzy.

Fuzziness of the component classes only at the level of the attribute values does not influence the created aggregation class. If there is fuzziness of the component classes at instance/schema level, namely, fuzziness at the second level, however, an additional attribute must be appended into the created aggregation class, indicating the aggregation degree of object instances. Fuzziness of the component classes at schema level, i.e., fuzziness at the first level, can result in the constraint rule in the created aggregation class, indicating the aggregation degree of the component classes.

8.4 Summary

Conceptual data models are generally able to capture complex object and semantic relationships at a high level of abstraction. At the level of data manipulation, logical database models are used for data storage, processing, and retrieval activities related to data management. Database modeling often starts with conceptual data modeling and then the created conceptual data models are transformed into logical database models.

This chapter has presented the transformations of the fuzzy ER and EER data models to the fuzzy relational and object-oriented database models, respectively. In addition, the IFO model was extended to the IF_2O and ExIFO models based on possibility distributions and similarity relations in (Ma, 2005) and (Yazici et al., 1999), respectively, and their mappings to the fuzzy and fuzzy nested relational database models were developed.

References

Abiteboul, S. and Hull, R. (1987), IFO: a formal semantic database model, ACM Transactions on Database Systems, 12 (4): 525-565.

Chaudhry, N. et al. , Moyne, J. and Rundensteiner, E. A. (1999), An extended database design methodology for uncertain data management, Information Sciences, 121: 83-112.

Fong, J. (1995), Mapping extended entity-relationship model to object modeling technique, SIGMOD Record, 24 (3): 18-22.

Halpin, T. A. (2002), Metaschemas for ER, ORM and UML data models: a comparison, Journal of Database Management, 13 (2): 20-30.

Ma, Z. M. (2005), A conceptual design methodology for fuzzy relational databases, Journal of Database Management, 16 (2): 66-83.

Poncelet, P., Teisseire, M., Cicchetti, R. and Lakhal, L. (1993), Towards a formal approach for object database design, Proceedings of the 19th International Conference on Very Large Data Bases, 278-289.

Shoval, P. and Frumermann, I. (1994), OO and EER conceptual schemas: a comparison of use comprehension, Journal of Database Management, 5 (4): 28-38.

Teorey, T., Yang, D. and Fry, J. (1986), A logical design methodology for relational databases using the extended entity-relationship model, ACM Computing Surveys, 18 (2): 197-222.

Yazici, A., Buckles, B. P. and Petry, F. E. (1999), Handling complex and uncertain information in the ExIFO and NF2 data models, IEEE Transactions on Fuzzy Systems, 7 (6): 659–676.

Zvieli, A. and Chen, P. P. (1986), Entity-relationship modeling and fuzzy databases, Proceedings of the 1986 IEEE International Conference on Data Engineering, 320-327.

9 Relational and Nested Relational Database Implementations of the Fuzzy IDEF1X and EXPRESS-G Models

9.1 Introduction

Information systems have become the nerve center of current computer-based industrial applications and engineering information modeling is thus essential. Viewed from database systems, engineering information modeling can be identified at two levels: conceptual data modeling and logical database modeling just like classical database modeling (Ma, 2005a). Also database modeling of engineering information generally starts from the conceptual data models and then the developed conceptual data models are mapped into the logical database models. For example, the EXPRESS data model was translated into relational databases (Krebs and Decker, 1995) and object-oriented databases in (Goh *et*., 1997), respectively. It should be noticed that the conceptual data models for engineering information modeling here are generally constructed by using IDEF1X or EXPRESS/EXPRESS-G in addition to ER/EER or UML (Sanderson and Spooner, 1993). The reason why IDEF1X and EXPRESS/EXPRESS-G rather than ER/EER and UML are extensively adopted in engineering information modeling is that they are more suitable for some industrial applications.

In order to represent fuzzy information in conceptual data models, IDEF1x and EXPRESS/EXPRESS-G as well as ER/EER and UML have been extended. Also the fuzzy relational, fuzzy nested relational and fuzzy object-oriented databases have been developed. Chapter 8 has investigated the conceptual design of the fuzzy relational databases and the fuzzy object-oriented databases by using the fuzzy ER model and the fuzzy EER model, respectively. In this chapter, we will discuss the issues how to transform the fuzzy IDEF1X data model and the fuzzy EXPRESS-G data model into the fuzzy relational database schema and the fuzzy nested relational database schema, respectively.

Z. Ma: *Fuzzy Database Modeling of Imprecise and Uncertain Engineering Information*, StudFuzz **195**, 167–177 (2006)
www.springerlink.com

9.2 Transformation from the Fuzzy IDEF1X Model to the Fuzzy Relational Database Model

The NF^2 data model is useful in engineering data modeling due to its capacity of modeling complex objects with hierarchical structure, which are very common in engineering areas.

9.2.1 The Fuzzy Relational Database Support to the Fuzzy IDEF1X Model

First of all, let us focus on relational databases support for IDEF1X model. The entity instances are identified by their primary key in IDEF1X data model. Entity instances can be represented as tuples in relational databases, where the tuples are identified by their key. When an entity type is mapped into a relation, the primary key of the entity can be viewed as the key of the tuples and the non-primary attributes of the entity become the common attributes of the tuples. So IDEF1X data model must contain a primary key. It should be noticed that, however, when we are mapping an identifier-dependent entity into a relation, the key of the tuples comes from the foreign key of the entity, which is essentially the primary key of the parent entity of this identifier-dependent entity. As to a non-identifying relationship, it is mapped into a relation, where the primary keys of two entities that are involved in this non-identifying relationship become the attributes of the mapped relation.

Second of all, let us focus on fuzzy relational databases support for the fuzzy IDEF1X model. In this case, the relational databases support for IDEF1X model is still available. In addition, fuzzy entity instances in the fuzzy IDEF1X model can be represented as tuples with degrees of membership in fuzzy relational databases. And fuzzy attribute values in fuzzy IDEF1X model can become fuzzy attribute values of tuples in fuzzy relational databases. As to a non-identifying relationship with the second level of fuzziness, it is mapped into a relation with membership degree attribute, where the primary keys of two entities that are involved in this non-identifying relationship become the attributes of the mapped relation. It should be pointed out that, although fuzzy relational databases can mainly support the fuzzy IDEF1X model, it couldn't support the meta-level fuzziness modeling in the fuzzy IDEF1X model because of the modelling capability of relational databases. The meta-level fuzziness in the fuzzy IDEF1X model includes fuzzy entities, fuzzy attributes, and the first level of fuzziness in connection relationships. Without considering the metalevel fuzziness in fuzzy IDEF1X model, in the following, we give a formal approach to transform a fuzzy IDEF1X schema to a fuzzy relational schema.

9.2.2 Transformation of Entities

Entities in the fuzzy IDEF1X model are generally transformed to relations in fuzzy relational databases and the attributes of entities may be considered as that of the corresponding relations directly. The primary keys (identifier-independent entities) or foreign keys (identifier-dependent entities) of entities in the fuzzy IDEF1X model become the key of the mapped relations. Two kinds of entities in the fuzzy IDEF1X model can be distinguished as follows.
- Entities without any fuzziness
- Entities with fuzzy entity instances

The first kind of entities can be directly transformed into relations. For the third kind of entities, however, an additional attribute as a non-key attribute, called membership degree attribute and denoted by pD, should be added to the relations transformed from the corresponding entities, which is used to denote the possibility that the tuples belong to the relation.

Figure 9.1 shows the transformation of entities.

Old-staff/1

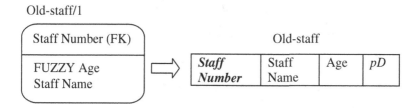

Fig. 9.1. Transformations of entities

9.2.3 Transformation of Connection Relationships

Non-identifying relationship of connection relationship in IDEF1X model should be mapped into an association relation, which attributes serve as a group of pointer to combine an explicit reference from one tuple to another tuple. Considering the constraint of cardinality, such attributes in two associated tuples can be single-valued one or multivalued one.

Let entity E_1 with attributes $\{K_1, A_1, ..., A_m\}$ and entity E_2 with attributes $\{K_2, B_1, ..., B_n\}$ be connected with non-identifying relationship R, where K_1 and K_2 are, respectively, key attributes of E_1 and E_2, and R may be one-to-one, one-to-many, or many-to-many relationship. In addition to the transformation process given in entity transformation discussion before, the relationship R is mapped into a relational schema with attribute set $\{K_1, K_2\}$, where K_1 and K_2 are key attributes. If the constraint of cardinality is a one-to-many relationship, i.e., R is a one-to-many relationship from E_1 to E_2, K_2 is a set-valued attribute in r_1 created by E_1, and K_1 is a single-valued attribute in r_2 created by E_2.

Considering the fuzziness in the fuzzy IDEF1X model, we distinguish two kinds of non-identifying relationships as follows.

- Relationships without any fuzziness;
- Relationships with the fuzziness at the second level;

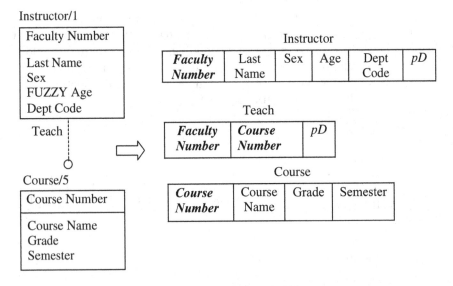

Fig. 9.2. Transformations of non-identifying relationships

For the first kind of relationships, the transformation can be conducted according to the rules given above. For the second kind of relationships, additional attributes denoting the possibility that the relationship instances belong to the relation should be added into the relation created by the relationship.

For an identifying relationship in connection relationship, the child entity must be an identifier-dependent entity. The association relation of the child entity with the parent entity is identified via the same primary key (the primary key of the child entity is essentially the foreign key, which inherits the primary key of the parent entity). Therefore, not being the same as the transaction of non-identifying relationship, it is not necessary to transform identifying relationships into relations.

Figure 9.2 shows the transformation of non-identifying relationships.

As to non-specific relationships, the transformation processing is the same as that of connection relationships.

9.2.4 Transformation of Categorization Relationships

In categorization relationships, the primary key of the category entities is inherited from that of the generic entity, so the category entities are identifier-dependent. The relationship between the generic entity and each category entity can be viewed as an identifying relationship. The association relation of each category entity with the generic entity is identified via the same primary key. So it is unnecessary to transform categorization relationships. Now we focus on how to transform the generic entity and the category entities.

The relationship between the generic entity and the category entities is essentially the specialization relationship between superclass and subclass. In general, the above-mentioned basic transformation rules for entities can be used. That is, entities are mapped into relations, and the attributes of the entities are mapped into the attributes of the mapped relations. In addition, if these entities are fuzzy ones with the fuzziness at instance/schema level, an additional attribute (membership degree attribute) pD should be added. However, the process of the primary keys in the categorization relationship transformation is different.

Let S be a generic entity with attributes named K, A_1, A_2, ..., and A_n, where K is its key. Let a category entity S_1 with attributes named A_{11}, A_{12}, ..., and A_{1k} and a category entity S_2 with attributes named A_{21}, A_{22}, ..., and A_{2m} be subclasses of S. Since S_1 and S_2 are subclasses of S, there are no keys in S_1 and S_2. At this point, S is mapped into the relational schema {K, A_1, A_2, ..., A_n}, and S_1 and S_2 are mapped into schemas {K, A_{11}, A_{12}, ..., A_{1k}} and {K, A_{21}, A_{22}, ..., A_{2m}}, respectively.

Figure 9.3 shows the transformation of specialization. Generic entity *Engine* is mapped into relation *Engine* with attributes *ID-Number* and *Model*. As to two categorization entities *Plane Engine* and *Car Engine*, they are mapped into relations *Plane Engine* with attributes *Name* and *Size* and relations *Car Engine* with attributes *Designer* and *Rate*, respectively. But categorization entities *Plane Engine* and *Car Engine* are subclasses of generic entity *Engine* and they do not have key. Therefore, the key *ID-Number* in generic entity *Engine* is added to relations *Plane Engine* and *Car Engine*, respectively.

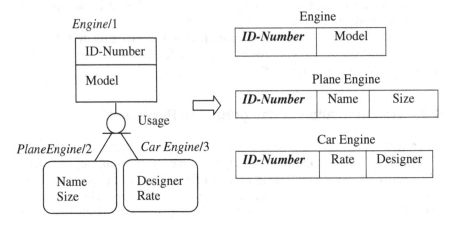

Fig. 9.3. Transformation of specialization

9.3 Transformation from the Fuzzy EXPRESS-G Model to the Fuzzy Nested Relational Database Model

The NF^2 data model is useful in engineering data modeling due to its capacity of modeling complex objects with hierarchical structure, which are very common in engineering areas. Let's look at the Instance-As-Type (IAT) problem proposed in (Erens, Mckay and Bloor, 1994; Li, Zhang and Tso, 2000). The IAT means that an object appears as a type in one information base but also as an instance in another information base at the same time. The IAT problems can result in the more difficulty and cost in the maintenance of information consistency. So we must resolve them in product data modeling, or update anomaly occur. The NF^2 data model can avoid the IAT problems naturally.

9.3.1 Nested Relational Database Support for EXPRESS Model

The entity instances are identified by their unique identifiers in EXPRESS information model. The entity identifiers are just like the keys in (nested) relational databases but they are different. The keys are the component parts of information content whereas the entity identifiers are not. We can view an entity as a database relation and view the instances of the entity as the tuples of the database relation. When we would like to represent entity instances in relational databases, we have to solve the problem how to identify entity instances in relational databases. In other words, we must indicate keys of the tuples originated from entity instances in the relational databases. As we know, in EXPRESS information, there are attributes with UNIQUE constraints. When an entity is mapped into a relation and each entity instance is mapped into a tuple, it is clear that such attributes can be viewed as the key of the tuples to identify instances. Therefore, EXPRESS information model must at least contain such an attribute with UNIQUE constraints when relational databases are used to model EXPRESS information model.

In EXPRESS, there is the entity which attributes are other entities, called *complex entities*. Complex entities and subtype/supertype entities in EXPRESS information model can be implemented in relational databases via the reference relationships between relations. It is clear that, based on such organization principles, the objects related in structural relationships one another are represented in separate relational databases. In order to obtain some information, one may have to query multiple relational databases by using join operations. Besides, it is very hard to have a complete and clear picture about information model from the query answers. Relational databases obviously are not suitable to support EXPRESS information model. The reason is because of the restriction of first normal form (1NF) in traditional relational databases. Nested relational databases can solve the problems above very well.

9.3.2 Formal Mapping

Generally speaking, an entity in EXPRESS-G can be mapped into a relational schema and the role names in the entity mapped into the attributes of the relational schema. But the following problems must be solved when mapping.

- How to represent the subtype/supertype relationship in the fuzzy NF^2 databases.
- How to model the fuzziness of the EXPRESS-G data model in the fuzzy NF^2 databases.
- Data type transformation.

It has been claimed above that the fuzziness in EXPRESS-G can be classified into three levels. The second level and the third level of fuzziness, namely, the fuzziness at the level of instance/entity and the fuzziness at the level of attribute value, can be represented in fuzzy NF^2 databases. Relational database models and nested relational database models only focus on instance modeling and their meta-structures are implicitly represented in the schemas. So the fuzziness at the level of entity and attribute cannot be modeled in fuzzy NF^2 databases due to the limitation of NF^2 databases in meta-data modeling.

The following three kinds of entities can be identified in EXPRESS-G model.

- Member entities. A member entity is the entity that comprises other entities as a component part or that is the underlying types of enumeration and select types.
- Subtype entities. A subtype entity is the entity that is in supertype/subtype relationships and is the subtype entity of the supertype entity/entities.
- Root entities. A root entity is neither a subtype entity nor a member entity.

For the entities mentioned above, we have different transformation strategies. First, let us look at the root entities. Each root entity is mapped into a fuzzy NF^2 relational schema and its role names become the simple attributes or structured attributes of the relation, depending on their data types.

- A role name with simple data type is mapped to a simple attribute.
- A role name with entity type is mapped to a structured attribute.
- A role name with defined type can be mapped to a simple attribute or a structured attribute, depending on the structure of the defined type.
- A role name with enumeration type is mapped to a simple attribute. Here, assume that NF2 databases support such attribute domain that users define themselves.
- A role name with select type is mapped to a structured attribute.

It is generally true that the transformations above should be processed recursively because an entity type, a defined type, or a select type may contain some of the above components by itself.

It should be noted that the fuzziness of values of simple attributes in EXPRESS-G information model is represented in the attribute values of NF^2 databases. For fuzziness of entity instances, the special attribute "pM" must be included in the structured attributes or the relational schema to indicate the membership degree of entity instances.

Second, let us look at the subtype entities. Following similar methods as described above, a subtype entity can be mapped into a fuzzy NF^2 schema.

It should be noted that, however, the key attributes of all the supertypes must be replicated in the subtype.

A member entity is not mapped to a class but a complex attribute of another class that is composed of the member entities. But the fuzziness in the member entity can be handled according to the same principles as the common entity.

Consider the fuzzy EXPRESS-G model shown in Figure 9.4. Following the formal rules given above, we map this fuzzy EXPRESS-G model into the fuzzy nested relational database in Table 9-1. Note that attribute "0.5/Thickness" in Figure 9.4 cannot be mapped into the fuzzy nested relational database due to its first level of fuzziness.

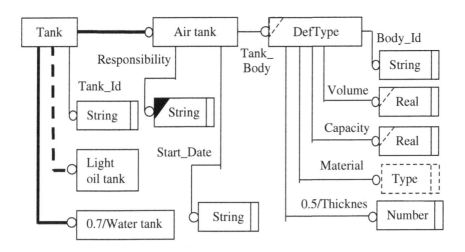

Fig. 9.4. A fuzzy EXPRESS-G data model

Table 9.1. Air tank relation

Tank_ ID	Tank_body				Start_ Date	Re- sponsi- bility
	Body_ID	Material	Volume	Capacity		
TA1	BO01	Alloy	about 2.5e+03	about 1.0e+06	01/12/99	John
TA2	BO02	Steel	about 2.5e+04	about 1.0e+07	28/03/00	{Tom, Mary}

It should be noted that we do not discuss the mapping of data types in fuzzy EXPRESS-G model. We assume that fuzzy nested relational databases support the data types in fuzzy EXPRESS-G model. In fact, the data types supported by different database products vary. More and more data types are supported by some latest release of database management systems. Our focus here is on mapping the entities and the attributes associated with entities in fuzzy EXPRESS-G model. We have identified all three kinds of entities in fuzzy EXPRESS-G model and given the mapping methods that map fuzzy entities and attributes into fuzzy nested relational databases. So the mapping methods given in the chapter can be used to solve the problem of fuzzy engineering data model transformations.

9.4 Summary

This chapter has presented the transformations of the fuzzy IDEF1X data model and the fuzzy EXPRESS-G data model to the fuzzy relational database schema and the fuzzy nested relational database schema, respectively. Also in (Ma, 2005b), the transformation of the fuzzy EXPRESS-G into the fuzzy relational database schema was presented.

It should be noticed that EXPRESS-G is only a subset of the full language of EXPRESS. Clearly it is necessary to implement the fuzzy EXPRESS data model in the fuzzy databases.

References

Erens, F., Mckay, A. and Bloor, S. (1994), Product modeling using multiple levels of abstract: instances as types, Computers in Industry, 24: 17-28.

Goh, A., Hui, S. C., Song, B. and Wang, F. Y. (1997), A STEP/EXPRESS to object-oriented databases translator, International Journal of Computer Applications in Technology, 10 (1-2): 90-96.

Krebs, T. and Decker, J. (1995), Translating EXPRESS models to the extended relational database management system POSTGRES, Proceedings of EUG '95 —The Fifth EXPRESS Users Group Conference.

Li, Q., Zhang, W. J. and Tso, S. K. (2000), Generalization of strategies for product data modeling with special reference to Instance-As-Type problem, 41 (1): 25-34.

Ma, Z. M. (2005a), Database modeling of engineering information: needs and constructions, Industrial Management and Data Systems, 105 (7): 900-918.

Ma, Z. M. (2005b), Modeling imprecise and uncertain engineering information in EXPRESS-G and relational databases, Proceedings of the 2005 ASME Design Engineering Technical Conference and Computers and Information in Engineering Conference (CD).

Sanderson, D. and Spooner, D. (1993), Mapping between EXPRESS and traditional DBMS models, Proceedings of EUG'93 — The Third EXPRESS Users Group Conference.

10 Object-Oriented Database Implementation of the Fuzzy EXPRESS Model

10.1 Introduction

The EXPRESS data modeling language is used to describe a product data model with activities covering the whole product life cycle. Based on such a product data model, product data can be exchanged and shared among different applications. This is the goal of the STEP standard. Generally speaking, the application of STEP is mainly concerned with two aspects: the establishment of the product information model that represents product data according to information requirements in an application environment and the integrated resources in STEP; and the manipulation and management of product data in the product information model. Both of these aspects are related to STEP implementation. Among the four levels of STEP implementation technology (i.e., file exchange, working form, database, and knowledge base), database implementation is currently receiving increasing attention.

In order to implement STEP with databases, the following tasks must be performed:

- Define the database structures from EXPRESS information.
- Provide SDAI access to the databases.

Here SDAI (Standard Data Access Interface) is an access protocol, i.e., a STEP API (application program interface), for the databases defined by EXPRESS. SDAI applications can be used to access the EXPRESS information model. But the specifications of SDAI operations are only given in STEP Part 23 and STEP Part 24. The implementations of these operations are empty. Their implementations must be performed by the specified binding language. Utilizing different database models, some mapping operations have been developed in the literature to map EXPRESS to databases. These database models include traditional databases such as network, hierarchical, and (nested) relational databases. The mappings of EXPRESS to the relational model have been shown in (Eggers, 1988;

Z. Ma: *Fuzzy Database Modeling of Imprecise and Uncertain Engineering Information*,
StudFuzz **195**, 179–204 (2006)
www.springerlink.com

Raghavan, 1992). A mapping to the Postgres extended relational model has also been reported in (Krebs and Decker, 1995). In addition, the mappings between EXPRESS and the network, hierarchical, and relational models are described in (Sanderson and Spooner, 1993). Since EXPRESS is semantically richer than traditional database models (Sanderson and Spooner, 1993), the mappings of EXPRESS to the object-oriented database model have been previously investigated. Some object-oriented database systems use programming languages as a DDL (Data Definition Language). Work on C++ mappings has also been done (Hardwick, 1991; Goh et al., 1994; Goh et al., 1997). Additional work on programming language mappings has been done by STEP WG11 during the development of the SDAI bindings (ISO TC184/SC4 WG7 N393, 1995; ISO TC184/SC4 WG7 N394, 1995). A general mechanism for identifying information loss between data models has been shown in (Sanderson, 1995). Compared with the mapping of EXPRESS to databases, limited work has been done on SDAI database implementations at present. SDAI implemented in object-oriented databases is reported in (Goh *et al.*, 1994) and some work with ObjectStore is described in (Herbst, 1994). A number of systems have been reported in (Krebs and Lührsen, 1995). It should be noted that, however, the STEP implementation with databases proposed in the literature is designed under the assumption that the data/information stored in the EXPRESS model is precise and the databases are crisp. In fact, these assumptions are often not valid for many real-world applications.

Since the fuzzy EXPRESS information models have been proposed, how to implement such models in database systems and reach the goal of STEP is the main issue of this chapter. The formal methods for mapping the fuzzy EXPRESS information models to the fuzzy object-oriented database schema will be developed. According to the feature of imperfect information models, the requirements of SDAI functions will then be investigated to manipulate the EXPRESS-defined data in the databases. Depending on object-oriented database platform, the implementation algorithms of these SDAI functions will be developed.

10.2   STEP and Implementation

The Standard for Exchange of Product Model Data (STEP) was born in 1983, when the International Standards Organization (ISO) formed the TC184/SC4 committee. STEP is an international standard that provides specifications for the representation and exchange of product information. Therefore, STEP is valuable for product databases, data exchange and concurrent engineering (Warthen, 1992). At present, the STEP effort is still very active. Many portions of STEP have been published as international standards, but many more are still under development.

10.2.1 Structure of STEP and Information Models

In a product's entire life cycle, many engineering activities are involved, including design to analysis, manufacture, quality control testing, inspection, and product support functions. In order to contain enough information in a product data model, STEP must cover geometry, topology, tolerances, relationships, attributes, assemblies, configuration and more. Therefore, STEP has been divided into a multiple-part standard. These parts cover general areas such testing procedures, file formats, and programming interfaces, as well as industry-specific information. STEP is extendable. Industry experts use EXPRESS to detail the exact set of information required for describing products of that industry. The *Application Protocols* form the bulk of the standard, and are the basis for STEP product data exchange.

Infrastructure	Information Models
Description Methods #11 EXPRESS #12 EXPRESS-I	**Application Protocols** #201 Explicit Drafting #202 Associate Drafting #203 Configure Control Design …
Implementation Methods #21 Physical File #22 SDAI Operations #23 SDAI C++ #24 SDAI C	**Application Integrated Resources** #101 Drafting #102 Ship Structures … Generic Integrated Resources #41 Miscellaneous #42 Geometry & Topology #43 Features …
Conformance Testing #31 General Concepts #32 Test Lab Requirements #33 Abstract Test Suites …	

Fig. 10.1. Structure of STEP

Table 10.1. STEP application protocols

Part 201 Explicit Draughting

Part 202 Associative Draughting

Part 203 Configuration Controlled Design

Part 204 Mechanical Design Using Boundary Representation

Part 205 Mechanical Design Using Surface Representation

Part 206 Mechanical Design Using Wireframe Representation

Part 207 Sheet Metal Dies and Blocks

Part 208 Life Cycle Product Change Process

Part 209 Design Through Analysis of Composite and Metallic Structures

Part 210 Electronic Printed Circuit Assembly, Design and Manufacturing

Part 211 Electronics Test Diagnostics and Remanufacture

Part 212 Electrotechnical Plants

Part 213 Numerical Control Process Plans for Machined Parts

Part 214 Core Data for Automotive Mechanical Design Processes

Part 215 Ship Arrangement

Part 216 Ship Moulded Forms

Part 217 Ship Piping

Part 218 Ship Structures

Part 219 Dimensional Inspection Process Planning for CMMs

Part 220 Printed Circuit Assembly Manufacturing Planning

Part 221 Functional Data and Schematic Representation for Process Plans

Part 222 Design Engineering to Manufacturing for Composite Structures

Part 223 Exchange of Design and Manufacturing DPD for Composites

Part 224 Mechanical Product Definition for Process Planning

Part 225 Structural Building Elements Using Explicit Shape Rep

Part 226 Shipbuilding Mechanical Systems

STEP is comprised of the infrastructure part and the information model part. The infrastructure part includes the Description Methods (EXPRESS) and Implementation Methods (file and programming interface). The information model part includes the industry-specific parts (application protocols). The infrastructure part has been separated from the information models part. Most of the infrastructure is complete, but the industry-specific parts are open-ended. Application protocols are available for mechanical and electrical products, and are under construction for composite materials, sheet metal dies, automotive design and manufacturing, shipbuilding, process plants, and others. In the long run, many industries need to develop their own standard application protocols. The structure of the STEP is shown in Figure 10.1.

Three kinds of STEP information models can be identified, namely, Application Protocols (*APs*), Integrated Resources (*IRs*), and Application Integrated Constructs (*AICs*). *APs* are built from general information models called Integrated Resources. In addition, STEP defines collections of common definitions that can be shared between Application Protocols. These Application Integrated Constructs are important when using data defined by several *APs*.

Application Protocols (APs)

An open-ended number of Application Protocols (*APs*) are defined in the STEP standard for industry-specific product data exchange. These *APs* are formal documents that cover a set of activities in the life cycle of a product. Every *AP* defines a set of activities, information requirements within this scope, and a formal schema for these requirements that is tied to an integrated product model shared between all *APs*. The STEP application protocols are designated as the 200-series documents. Each AP covers a portion of a product lifecycle. Some of these have reached international standard status while others are still under development. A list of current *APs* are shown in Table 10.1.

Among these *APs*, each one covers a set of activities in the lifecycle of a product. *APs* 202 through 209, for example, deal with aspects of the design and analysis of mechanical parts. *AP* 214 further narrows this scope to automotive parts. *APs* 210, 211, and 220 cover aspects of circuit board manufacture. The statement of this scope is called the Application Activity Model (*AAM*). The *AAM* is normally documented using *IDEF0* (IDEF, 2000) diagrams.

The next portion of an *AP* describes product information, called Application Reference Model (*ARM*), required for the activities. This model is concise and describes requirements in terms of basic Application Objects that a user of the *AP* information would be concerned with. The application objects can be described by *IDEF1X* (Kusiak *et al.*, 1997; IDEF, 2000), *EXPRESS-G, ER* (Chen, 1976), or *EER* diagrams.

Application objects can be grouped into subject areas called Units of Functionality (*UOF*). The *UOF*s describe a logically complete subset of information about some particular product aspect. *AP* 203 (ISO 10303-203, 1994), for example, contains 36 application objects, distributed among nine units of functionality. The *UOF*s are Authorization, Bill of Material (BOM), Design Information, Design Activity Control, Effectivity, End Item Identification, Part Identification, Shape, and Source Control.

Within each *UOF* are application objects that represent the information needed to describe that product aspect. For example, the Design Activity Control *UOF* tracks product modifications. The application objects in this *UOF* are Change Order, Change Request, Start Order, Start Request, Work Order, and Work Request.

Finally, an *AP* document contains a conceptual schema that describes the ARM in terms of a library of pre-existing definitions. This Application Interpreted Model (*AIM*) is always described with EXPRESS, and is based on the definitions from the integrated resources described in the next section. *AIM*s are not permitted to define new entities. They are only permitted to refine definitions already present in the integrated resources. This restriction prevents the same concepts from being modeled in different ways by different *AP*s.

Integrated Resources (IRs)

The Integrated Resources (*IR*s) are the cores of STEP. These conceptual schemas describe an integrated product model for all *AP*s. Among *IR*s, there are two kinds of *IR*s identified. They are the generic integrated resources (40-series documents) and the application integrated resources (100-series documents). The generic integrated resources describe very general characteristics of products across all industries and the application integrated resources refine the integrated resources down to the needs of a particular industry. A list of current *IR*s is shown in Table 10.2. Some of these have reached international standard status while others are still in various stages of development.

The resources vary with their level of detail. For example, Part 41 (ISO 10303-41, 1994) covers product identification. Since a product could be a camshaft or an office building, these definitions are very general and are normally refined by an AP or application integrated resource. On the other hand, Part 42 (ISO 10303-42, 1994) describes geometry, which is well-defined out of the context of any particular application, so this part is normally used without additional refinement.

Table 10.2. STEP integrated resources

Part 41 Product Description and Support
Part 42 Geometric and Topological Representation
Part 43 Representation Structures
Part 44 Product Structure Configuration
Part 45 Materials
Part 46 Visual Presentation
Part 47 Shape Tolerances
Part 48 Form Features
Part 49 Process Structure and Properties
Part 101 Draughting Resources
Part 102 Ship Structures
Part 103 Electrical/Electronics Connectivity
Part 104 Finite Element Analysis
Part 105 Kinematics

Application Integrated Constructs (AICs)

STEP recently introduced a construct for describing the interoperable segments of definitions shared by multiple *AP*s. The constructs, called Application Integrated Constructs (*AIC*s), are sets of refined definitions that must be used as a single unit, without any additional refinements.

10.2.2 STEP Implementation Methods and Levels

EXPRESS information models describe logical structures. These structures must be mapped to a software technology before they can be used. A model might be implemented using many data processing technologies and would remain relevant as new technologies appear.

STEP Implementation Methods

A number of implementation methods for exchanging and manipulating information described by application protocols have been defined in STEP. The first implementation method is STEP physical file exchange and the second implementation method is STEP Data Access Interface (SDAI).

STEP Physical File Exchange. In this implementation method, a straightforward ASCII file is used for exchanging the data sets in EXPRESS model. This exchange file format is described in (ISO 10303-21, 1994), which contains a header section with identifying information, as well as a data section. Data section in a STEP exchange file contains the information to be transferred.

STEP Data Access Interface (SDAI). This implementation method is to provide an access protocol for the databases defined by EXPRESS. The access protocol is called Standard Data Access Interface (SDAI). The goal of SDAI is to provide the users with uniform manipulation interfaces and reduce the cost of integrated product databases. The SDAI is comprised of several ISO standards documents. STEP Part 22 (ISO TC184/SC4 WG7 N392, 1995) contains a functional description of the SDAI operation. STEP Part 23 (ISO TC184/SC4 WG7 N393, 1995) and STEP Part 24 (ISO TC184/SC4 WG7 N394, 1995) describe how these operations are made available in the C++ and C language environments. Required by the applications, binding other languages such as CORBA, Java, and etc. are also being considered. Using SDAI operations, the SDAI applications can access EXPRESS information model. It should be noted that only the specifications of SDAI operations are given in STEP Part 23 and STEP Part 24. The implementations of these operations are empty. Their implementations must be performed by the specified binding language. In general, the SDAI language bindings can be classified into early binding and late binding.

STEP Implementation Levels

Based on the implementation methods above, four levels of implementation technology have been identified for EXPRESS models. These levels are shown in Figure 10.2.

Level Two: Working-Form. In addition to all features required by Level One, at this level, the ability to manipulate data based on the EXPRESS information model is necessary. A working-form application communicates with other applications using exchange files. However, when an application reads a file into the memory, the data is made available to the code in a form organized and described by the EXPRESS model. The SDAI has been developed as a standard API for working-form applications. The

SDAI functions allow programs to manipulate any product data defined by an EXPRESS model. Other programming bindings could also be used, as long as they are based upon the EXPRESS models.

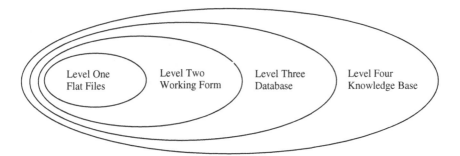

Fig. 10.2. STEP implementation levels

Level Three: Database. A Level Three implementation has all features required by Level Two, as well as the ability to work with data stored in a Database Management System (DBMS). An integrated product database may store data that covers many aspects of the engineering life cycle. Multiple applications can access the product data, and may take advantage of database features, such as query processing. The implementation should also be able to read and write exchange files and make product data available using either the SDAI or another API that presents data as structures defined by EXPRESS. A database implementation may support the validation of some EXPRESS constraints, but need not to support all of them. At present, limited work has been done on SDAI database implementations. Although several groups have looked at mapping EXPRESS-definitions into database structures, only a few have attempted to provide SDAI access to the structures. SDAI implemented in object-oriented databases is reported in (Goh *et al.*, 1994) and some work with ObjectStore is described in (Herbst, 1994). A number of systems have been reported in (Krebs and Lührsen, 1995).

Level Four: Knowledge Base. A Level Four implementation has the features of all lower level implementations, plus full support for EXPRESS constraint validation. A knowledge base system should read and write exchange files, make product data available to applications in structures defined by EXPRESS, work on data stored in a central database, and should be able to reason about the contents of the database. Knowledge base systems encode rules using techniques such as frames, semantic nets, and various logic systems, and then use inference techniques such as forward and backward chaining to reason about the contents of a database. Although some interesting preliminary work was done (Muller and Smith, 1993), knowledge base implementations do not exist.

In any case, EXPRESS information models describe logical structures that must be mapped to an implementation technology before use. The file and working-form implementations have the fewest requirements and are widely available. Database and knowledge base implementations have more requirements and are not common. In the following, ways to satisfy the requirements for database (Level Three) implementations are focused on.

10.2.3 STEP Database Implementation Process

In order to construct an engineering database around an EXPRESS information model, the following tasks must be performed.
- Define the database structures from EXPRESS information.
- Provide SDAI access to the database.

The second task requires several design decisions. First of all, SDAI is a STEP API (application program interface) for the data defined by EXPRESS. In order to transfer these data between database and applications, SDAI operations must be performed according to the specified binding language and database platform. As to how to transfer data between databases and applications, SDAI access architectures are involved. Three implementation architectures of SDAI in databases, which are File Upload/Download SDAI Binding, Cached SDAI Binding, and Direct SDAI Binding, have been identified and the performance of their implementations has been examined in (Loffredo, 1998).

An EXPRESS information model must be converted into schema definitions for the target database. This conversion requires a mapping from the EXPRESS language into the data definition language (DDL) of the target database system. An existing mapping can be used if applicable. But some database systems may require a new EXPRESS to DDL mapping. EXPRESS information models can challenge the capabilities of existing database systems.

In particular, the following features of EXPRESS may require encoding or other manipulations to preserve the original information within the native data model.
- Entities: Each entity instance, not requiring unique keys formed from attribute values, has an implicit identifier, which is used to store relationships. Database systems that rely on unique keys for identification will probably need to add additional identifier data to entities.

- Inheritance: The EXPRESS inheritance model includes single inheritance and multiple inheritances. It also supports AND/OR inheritance, where an instance may have a set of types. Since inheritance implies duplication of attributes between supertypes and subtypes, normalization may be needed for some database systems such as relational databases (Abiteboul *et al.*, 1995).
- Simple data types. There are seven simple data types in EXPRESS. New types can be defined by adding constraints to existing types. Encoding may be needed for some of the simple data types and defined types.
- Enumeration: Each set of enumerated values is in a separate name space. Not all database systems support enumeration as a primitive type.
- Selects: The EXPRESS select type is analogous to a strongly typed union and is used to group disjoint types. Selects can be formed from any number of base types, and can be nested to arbitrary depth. Very few database systems can support a union type.
- Aggregates: EXPRESS supports ordered and unordered aggregates formed of any base type and nested to arbitrary depth. These types of structures are only supported by non-first normal form databases. Even with these, some aggregate styles may need to be simulated.

A mapping from EXPRESS to a database schema should address each of these constructs. Depending upon the database, derived attributes, local and global constraints, unique, or inverse clauses may also be addressed.

10.3   Conversion of the Fuzzy EXPRESS Models to the Fuzzy Object-Oriented Databases

10.3.1 Object-Oriented Database Support for EXPRESS Data Model

Unlike the relational databases, there is no widely accepted definition as to what constitutes an object-oriented database, although an object-oriented database standard has been released by ODMG (2000). The ODMG group completed its work on object data management standards in 2001 and was disbanded. Not only is it true that not all features in one object-oriented database can be found in another, but the interpretation of similar features may also differ. However, some features are in common with object-oriented databases, including object identity, complex objects, encapsulation, types, and inheritance. EXPRESS is object-oriented by nature, which supports these common features in object-oriented databases. Therefore, there should be a more direct way of mapping the EXPRESS data model into object-oriented databases.

It should be noted that there is incompatibility between the EXPRESS data model and object-oriented databases. No widely accepted definition of an object-oriented database model results in the fact that there is no common set of incompatibilities between EXPRESS and object-oriented databases. Some possible incompatibilities illustrated in (Goh *et al.*, 1997) are listed in Table 10.3.

Table 10.3. Incompatibility between EXPRESS and OODBs

Data aggregation	Not all OODBs support the four kinds of aggregate data type in EXPRESS. Especially, BAG may not be supported.
SELECT data type	SELECT data type is not supported in many OODBs as it causes ambiguity in the class hierarchy. However, incomplete OODBs support such data type well.
Late binding	General polymorphism is expected by EXPRESS whereby dynamic binding can be achieved. Some OODBs permit only extension polymorphism that allows such dynamic binding only if there is a parent-child class relationship.
Type vs. class	EXPRESS perceives a class as a collection of instances upon which queries or rules can apply. Some OODBs are type-based systems, as opposed to class-bases systems.
Constraints	Many OODBs do not support a trigger mechanism or a constraint specification language. So EXPRESS constraints have to be programmed in a general purpose object-oriented programming language as class methods which are accessed privately by public methods. This hinders maintainability of the constraints.

10.3.2 Formal Transformation

The first level of fuzziness in the EXPRESS data model cannot be modeled in the nested relational databases because such database models have a weak ability to model the meta structures of the data models. In addition, the subtype/supertype relationship in the EXPRESS data model is not also modeled in the relational databases and the nested relational databases directly. All these problems are well resolved in fuzzy object-oriented databases because the fuzzy EXPRESS model and fuzzy object-oriented databases (FOODBs) are in fact well matched. In the fuzzy EXPRESS model, entities, rules, functions, and procedures are its main parts. The domain rules and local rules are related to attribute definitions. When the fuzzy EXPRESS model is mapped to the fuzzy object-oriented database, the entities are mapped to the classes and the domain rules can be viewed as the domain integrity constraints of the attributes in the FOODBs. Other rules as well as the functions and the procedures are all mapped to the methods in the FOODBs. In the following, let us focus on the transformation of entities, involving attributes, data types, subtype/supertype and different levels of fuzziness.

First let us focus on the entities that are not membership entities. For an entity in the EXPRESS model, if it is the entity with WITH ... DEGREE, then it is mapped to a class with WITH DEGREE OF. Otherwise, it is mapped to a common class. If the entity is a subtype of other entities, then in the created class, the superclasses created by the supertypes should be declared. Since the entity may have a different degree with respect to each of its supertypes, these superclasses also have the degrees themselves, which must be respectively indicated.

Attributes of an entity are mapped to attributes of the class created by the entity. If an attribute of the entity is with WITH ... DEGREE, then it is mapped to an attribute of the class, which is with WITH DEGREE OF. In addition, when there is the fuzziness in the level of entity instance/entity type in an entity and the entity is mapped to a class, an additional attribute pD should be added to the class created. The value of this attribute is used to denote the possibility that the corresponding instance belongs to the class. The derived attributes and inverse attributes in the EXPRESS model are mapped to the methods of the class created as constraints.

The following investigates how various data types in the EXPRESS model can be mapped to the FOODB. The focus is on simple data type, collection data type, entity type, defined type, enumeration type, and select type. The attributes with simple data type, crisp or fuzzy, in the EXPRESS model can directly be mapped to the simple attributes of the class created. The attributes with enumeration type, crisp or fuzzy, in the EXPRESS model are also mapped to the simple attributes of the class created. Here, assume that the FOODB also supports such attribute domain that the users define themselves. The attributes with defined type in the EXPRESS information model are mapped to the complex attributes of the class created. For the attributes with the collection data type in the EXPRESS model, if the underlying type is a simple data type, then they are mapped to the simple attributes of the class created with multiple values. If the underlying type is not a simple data type, the transformations of attribute are decided by the data type of the underlying type.

```
ENTITY C WITH 0.9 DEGREE
SUBTYPE OF (X WITH 0.8 DEGREE, Y WITH 0.7 DEGREE);
    A1: INTEGER WITH 0.8 DEGREE;
    A2: FUZZY REAL;
    A3: SET [1 : ?] OF REAL;
    A4: Enum;
    A5: Sele = FUZZY SELECT (0.3/REAL, 0.6/INTEGER, 0.4/STRING);
    A6: Enti;
    pD: NUMBER;
    DERIVED
    B: FUZZY REAL := A1 ** 2 – 2 * A1 * A2 + A2 ** 2;
END_ENTITY;
    TYPE Enum = ENUMERATION (e1, e2, e3);
    END_TYPE;
    ENTITY Enti
    D1: REAL;
    D2: FUZZY INTEGER;
    END_TYPE;
```

Fig. 10.3. Fuzzy EXPRESS data model

An attribute with select type in the EXPRESS model can be viewed as a group of attributes that have the same attribute name but a different data type. The number of the group of attributes and the data type of each attribute are determined by the content of this select type. Each attribute in it has a value of a membership degree, being greater than 0 and less than or equal to 1 for fuzzy select type but equal to 1 for crisp select type, which indicates the possibility that the attribute takes the corresponding data type. Therefore, according to the content of the select type, an attribute with select type in the EXPRESS model is mapped to multiple attributes with WITH DEGREE OF in the class created.

Finally, let us look at the member entities. A member entity is essentially an entity type. Such an entity is not mapped to a class but to a complex attribute of another class that is composed of the member entities. But the fuzziness of the member entity is processed according to the same principles as the common entity.

Based on the mapping methodologies above, an example is given. The EXPRESS model in Figure 10.3 is mapped to an FOODB model in Figure 10.4.

```
CLASS C WITH DEGREE 0.6
INHERITS X WITH DEGREE OF 0.8;
INHERITS Y WITH DEGREE OF 0.7;
ATTRIBUTES
    A1: TYPE OF Integer WITH DEGREE OF 0.8;
    A2: FUZZY TYPE OF Real;
    A3: DOMAIN set values TYPE OF Real;
    A4: DOMAIN {e1, e2, e3} single value;
    A5: TYPE OF Real WITH DEGREE OF 0.3;
    A5: TYPE OF Integer WITH DEGREE OF 0.6;
    A5: TYPE OF String WITH DEGREE OF 0.4;
    A6:
        D1: TYPE OF Real;
        D2: FUZZY TYPE OF Integer;
    pD: DOMAIN [0, 1] TYPE OF number;
METHODS
    B := A1 ** 2 – 2 * A1 * A2 + A2 ** 2;
    ...
END
```

Fig. 10.4. FOODB model created via fuzzy EXPRESS model

10.4   Function Requirements for Manipulating the Fuzzy EXPRESS Model

When the EXPRESS model is mapped to databases, users will use databases. Viewed from database systems, there are different levels of users of a database. In general, they can be classified into system programmer, database administrator, and application users (Abiteboul *et al.*, 1995). A system programmer is responsible for the physical level whereas the database administrator defines and maintains the database schema. The application users access and manipulate the data.

SDAI can be viewed as the data access interface. Therefore, SDAI falls into the category of the application users. The requirements of SDAI functions are decided by the requirements of the application users. Under the fuzzy information environment, the requirements needed for manipulating the fuzzy EXPRESS information model must consider the fuzzy information processing. However SDAI itself is in a state of evolution. Considering the enormity of the task and the difficulty of achieving agreement as to what functions are to be included and the viability of implementing the suggestions, only some basic requirements are focused on here, including data query, data update, structure query, and validation.

10.4.1 Data Query Requirements

The functions of a data query should support the retrieval of information that involves the access of information relating to a single entity instance. As we know, there is a set of entity instances corresponding to a given entity type and an entity instance is composed of attribute values. Therefore, the functions of a data query include the following operations:

- The retrieval of specific attributes or all attributes of a given entity instance. The result of the operation is the attribute value(s).
- The retrieval of a restricted number or all instances of the entity type.

For the first situation, users are required to provide the identifier of the entity instance so that the entity instance can be located. The corresponding attribute values are then obtained. For the second situation, the users should provide the condition that entity instances should satisfy. Such entity instances are in this way exported. If there is no condition provided, all entity instances of the entity type are obtained. Therefore, the operation in the first situation can be viewed as a specific case of the operation in the second situation. In addition, users may be only interested in specific attributes for the operation result in the second situation.

Based on the discussion above, a generic data query should be that for a given entity type, all the entity instances that satisfy the retrieval condition are obtained and the present values of identified attributes of these instances are returned. A formal function representing such a data query is as follows:

get_data (name_entity-type, [query_condition], [attribute_list]),

where the second and third parameters can be omitted. At this point, the default of the second parameter means that all instances of the entity type will be obtained. The default of the third parameter means that all attribute values of the instances satisfying the condition will be provided. Note that the first parameter is not optional.

In a query condition, the identifier of an entity instance can be directly provided. Generally speaking, however, a query condition is a Boolean expression, which is composed of some relational expressions as well as logical operators (AND, OR, NOT, and XOR). The relational expression has such a basic form as

attribute_name θ constant

Here, θ represents the relational operators in EXPRESS. In order to implement flexible query processing, the constant in a relational expression may be a partial value or a fuzzy value. The query condition containing such relational expressions above is called a flexible query condition. In general, a flexible query may return multiple instances that indefinitely satisfy the condition and have different degree with respect to the condition.

Example: Consider an entity type *Sofa* with attributes *Color*, *Length*, *Width*, and *Height* as well as the membership degree attribute *pD*. A query **get_data** (*Sofa, Length = about 2000 mm*, {*Color, Length, Width*}) is made. Assume that one obtains the following three groups of answers: {*Color = Grey, Length = 2000 mm, Width = 600 mm*}, {*Color = Black, Length = 1900 mm, Width = 580 mm*}, and {*Color = Brown, Length = 2200 mm, Width = 670 mm*}. It can be seen that all three groups of answers satisfy the flexible query condition. It is clear that the first group of answers has the greatest possibility of satisfying the condition. If the query condition is not flexible, say *Length = 2000 mm*, only the first group of answers is returned because the last two groups of answers do not satisfy the condition.

Since the EXPRESS model may be imperfect, the query answers under precise query conditions or flexible query conditions may return imperfect attribute values. In order to satisfy the requirements of an imperfect information query, the relational operators in relational expressions of query conditions should be the operators defined in Section 4.3. In addition, an entity instance belongs to the entity type with a membership degree. There is an additional attribute in the entity type declaration to represent the membership degree of instances. Such an attribute may appear in the relational expressions and the attribute list of query output. In other words, users can retrieve the instances satisfying the given membership degree and display the values of such a membership degree.

Multiform and flexible data query functions are crucial. This is especially true for querying the fuzzy EXPRESS information model. In the following section, the refinement of incomplete query answers will further be discussed.

10.4.2 Data Update Requirements

The functions of data update should support entity instances to be created, deleted, and modified. In the following, these three operations are discussed.

Data Creation

Data creation means that entity instances are created and are inserted into the set of instances of the entity type. Generally, users should assign values to all the attributes, including the membership degree attribute of the entity type under an incomplete information environment. Note that it is also permitted to create a partial instance, i.e., there are only some attributes or even no attribute to assign values to. At this point, the attributes to which no values are assigned take null values. A uniform form is used to represent data creation:

create_data (name_entity-type [, attribute-value_list])

Here, the second parameter can be omitted. At this point, an empty instance is created. The attribute value list is composed of elements with the form *attribute_name = attribute_value*. The attribute name and the attribute value must have a compatible data type and the attribute value may be fuzzy.

Example: Consider the above-mentioned entity type *Sofa* again. Now one can create two instances utilizing **create_data** (*Sofa*, {*Color = Green, Length = about 2100 mm, Width = 580 mm, Height = 760mm, pD = 0.8*}) and **create_data** (*Sofa*).

It should be noted that there is a subtle difference in partial creation between relational databases (RDBs) and object-oriented database systems. Entity instances in OODB are identified by the identifiers whereas they are identified by their key values in RDBs. Therefore, when partial creation functions are implemented in RDBs, the key attributes must have assigned values. A non-identified instance is not permitted in relational databases.

Data Deletion

Data deletion means that some entity instances are chosen and deleted from the set of instances of the entity type. Generally, users should provide the condition that the deleted instances should satisfy. Being similar to data query processing, the deleted instances are selected according to the given condition and are then deleted. Here the deletion condition, being the same as a query condition, can be flexible. Of course, full data deletion should be supported, which means that all instances are deleted. A formal representation of data deletion is as follows:

delete_data (name_entity-type [, deletion_condition])

Here, the second parameter can be omitted. At this point, all instances of the entity type are deleted. The deletion condition is the same as the query condition.

Example: Consider the entity type *Sofa* again. Now one performs the following deletion operation: **delete_data** (*Sofa*, {*Length* = *about 2100 mm*, AND *Width* = *670 mm*}).

Data Modification

Data modification means that some attribute values of entity instances are replaced with new attribute values. In order to perform such operations, users should provide two aspects of conditions: one is the condition that the modifying instances should satisfy before the modification; the other is the condition that the modified instances should satisfy after the modification. The first condition, denoted *condition_1*, is essentially viewed as a query condition. The second condition, denoted as *condition_2*, is composed of the elements that have the form as follows:

attribute_name = constant

Data modification is done through locating the instances that satisfy *condition_1*, and then the attribute values of these instances are replaced with the corresponding attribute values given in *condition_2*. It can be seen that the data modification is essentially performed through data deletion and data creation jointly. Note that the membership degree attribute of the entity type can appear in *condition_1* and *condition_2*. In addition, attribute names and attribute values in *condition_2* should have compatible data types. A formal representation of data modification is as follows:

modeify_data (name_entity-type, [condition_1], condition_2)

Note that the second parameter can be omitted. This means that all instances should be modified according to the *condition_2*. The third parameter cannot be omitted.

Example: Consider the entity type *Sofa* again. Now one performs the following modification operation: **modify_data** (*Sofa*, {*Width* = *670 mm*}, {*Length* = *2150 mm*, pD = *0.95*}).

It should be noted that for data creation and data modification, attribute names and attribute values given must have compatible data type. In addition, the data creation, data deletion, and data modification should also be subjected to the validation procedures stated in Section 10.4.4.

10.4.3 Structure Query Requirements

In the above two sections, the manipulation of attribute values of entity instances is discussed. Here an assumption is made that users know the entity description, namely, the entity structure. However, this is not always true. In addition, users may not know the schema description. Therefore, SDAI functions should provide some operations to enable querying into the schema.

The functions of a structure query include the following operations:

- The retrieval of entity type of a given schema. The result of the operation is the names of all entity types.
- The retrieval of attribute names and their definitions.

Schema Structure Query

Users are required to provide the name of the schema. The result of the operation returns the names of all entity types included in the schema. A formal representation of a schema structure query is as follows:

get_schema_definition (name_schema-type)

The one parameter cannot be omitted.

Entity Structure Query

Users are required to provide the name of the entity type. The result of the operation returns the names of all attributes included in the entity type. A formal representation of an entity structure query is as follows:

get_entity_definition (name_entity-type)

The one parameter cannot be omitted here either.

Example: For the entity type *Sofa* above, for example, one can perform the following entity structure query **get_entity_definition** (*Sofa*). Then one gets {*Color, Length, Width, Height, pD*}.

Attribute Definition Query

Users are required to provide the name of the entity type and a list of attribute names. The result of the operation returns the definitions of the given attributes in the entity type. A formal representation of attribute definition query is as follows:

get_attribute_definition (name_entity-type [, {attribute-name_list}])

Note that the second parameter can be omitted. This means that the definitions of all attributes should be returned.

Example: For the entity type *Sofa* above, for example, one can perform the following entity structure query: **get_attribute_definition** (*Sofa*, {*Color, Length, Height*}). Assume that {*Color = STRING, Length = FUZZY REAL, Height = REAL*} is obtained.

10.4.4 Data Validation Requirements

In the SDAI interface, simple error handling should be provided. It should be noted, however, that there are a variety of possible causes of failures. It may be difficult to recover from the errors. Since the onus is on the application to ensure that data integrity is preserved, validation is important. Data validation needs to support implicit checking to ensure that references made by the new instance exist. In addition, the data must conform to the schema definition and ensure that constraints declared in EXPRESS have been met. Fortunately, most database management systems support such checks before the data is made persistent.

10.5   SDAI Implementation on the Fuzzy Object-Oriented Databases

The function requirements given in the previous section can be viewed as the extension proposed in (ISO TC184/SC4 WG7 N23, 1991) to satisfy the need of manipulating the fuzzy EXPRESS information model. As we know, SDAI specifications (ISO TC184/SC4 WG7 N392, 1995) contain further clarification and elaboration of SDAI function requirements. In addition, language binding that gives a clear picture of the relationship between language binding and the specification is included in (ISO TC184/SC4 WG7 N393, 1995; ISO TC184/SC4 WG7 N394, 1995). It is required that each binding language must implement all functions given in the specification. Note that although the implementation formats of binding language are proposed in the specification, the contents of the functions are empty, which should be developed utilizing the special binding language according to database systems.

It should be noted that there are double difficulties met in the database implementation of SDAI, that is, not only are the SDAI specifications still in a state of evolution, but also the implementation of SDAI functions are product-related. Therefore, the algorithms for implementing the functions mentioned in Section 10.4, not being a complete, comprehensive library of routines, are developed in the following.

10.5.1 Data Query

In object-oriented databases, data query follows a procedure such that for a given class, the identifiers of the instances satisfying the specified conditions can be returned. Such identifiers can be transformed into the pointers to these instances. Access to each instance can then be realized. Here, the specified condition should be flexible in order to query fuzzy object-oriented databases. In addition, the access to instances can be made on attribute values.

The implementation algorithm of data query in object-oriented databases is given as follows.

Algorithm 10.1: SDAI_get_data (*class_name*, [*query_condition*], [*attribute_list*]);

Input: Class *class_name*, query condition *query_condition*, and attribute list *attribute_list*, where the last two parameters can be omitted;

Output: The values of the specified attributes of the specified instances;

Step 1: Find the class with name *class_name*;

Step 2: $T := \Phi$;

Step 3: If *query_condition* = NULL, then for each instance o of *class_name*, get its identifier *id_o* and run $T := T + \{id\_o\}$, and go to Step 5;

Step 4: For each instance o of *class_name*, if o satisfies *query_condition*, then get its identifier *id_o* and run $T := T + \{id\_o\}$;

Step 5: If *attribute_list* = NULL, then for each $t \in T$, get its pointer *pointer_t* and return the values of all attributes of the instance pointed by *pointer_t*, and go to Step 7;

Step 6: For each $t \in T$, get its pointer *pointer_t* and return the values of the given attributes of the instance pointed by *pointer_t*;

Step 7: Return (T).

10.5.2 Data Creation

To create a new instance of a class, a list of the attribute values is provided together with the class name. The class assigns an identifier for the new instance. In object-oriented databases, a completely partial creation is permitted, i.e., all attributes have no values. Partial creation refers to cases where some attributes do not have values. Just like the situation in relational databases, the new instance may contain incomplete information in attribute values and at the instance. In addition, it should be done to check the validation of the new instance.

The implementation algorithm of data creation in object-oriented databases is given as follows.

Algorithm 10.2: SDAI_create_data (*class_name* [, *attribute-value_list*]);

Input: Class *class_name* and attribute values list *attribute-value_list*, where the last parameter can be omitted;

Output: Return the identifier of the new instance in the class;

　　　　Step 1: Find the class with name *class_name*;

　　　　Step 2: If *attribute-value_list* ≠ NULL and it is not available, then return;

　　　　Step 3: Get a new pointer from *class_name*;

　　　　Step 4: Pass the attribute values to the pointer if there is any attribute values in the *attribute-value_list*;

　　　　Step 5: Get the identifier of the instance pointed by the pointer;

　　　　Step 6: Return (T).

10.5.3 Data Deletion

In order to delete instances from the specified class, the identifiers of the instances satisfying the specified condition can be returned. Such identifiers can be transformed into the pointers to these instances. Deletion of each instance can then be realized. Here, the specified condition should be flexible.

The implementation algorithm of data deletion in object-oriented databases is given as follows.

Algorithm 10.3: SDAI_delete_data (*class_name*, [*delete_condition*]);

Input: Class *class_name*, deletion condition *delete_condition*, where the last parameters can be omitted;

Output: Class *class_name* after the deletion;

　　　　Step 1: Find the class with name *class_name*;

　　　　Step 2: $T := \Phi$;

　　　　Step 3: If *delete_condition* = NULL, then for each instance *o* of *class_name*, get its identifier *id_o* and run $T := T + \{id\_o\}$, and go to Step 5;

　　　　Step 4: For each instance *o* of *class_name*, if *o* satisfies *delete_condition*, then get its identifier *id_o* and run $T := T + \{id\_o\}$;

　　　　Step 5: For each $t \in T$, get its pointer *pointer_t* and delete the values of all attributes of the instance pointed by *pointer_t*;

　　　　Step 6: Return (T).

10.5.4 Entity Structure Query

Data creation means that entity instances are created and are inserted into the set of instances of the entity type.

The implementation algorithm of entity definition in object-oriented databases is given as follows.

Algorithm 10.4: SDAI_get_entity_definition (*class_name* [, *property_list*]);

Input: Class *class_name* and property list *property_list*, where the last parameter can be omitted;

Output: The definitions of attributes in *property_list*;

Step 1:   Find the class with name *class_name*;

Step 2:   If *property_list* = NULL, then A := all properties of the class else A := *property_list*;

Step 3:   T : = Φ;

Step 4:   For each x ∈ A, get its attribute name *name_x* and type definition *type_x*, and then run T := T + {*name_x* : *type_x*};

Step 5:   Return (T).

10.6   Summary

This chapter has investigated the issues involved in implementation of the fuzzy EXPRESS information model on fuzzy object-oriented databases. Database implementation of EXPRESS information model involves information mapping from the EXPRESS data model into database structures and the development of SDAI operations. In the chapter, formal methods for mapping fuzzy EXPRESS information model to the fuzzy object-oriented databases have been developed. Taking into account the features of the imperfect EXPRESS information models, the requirements of SDAI functions have been investigated to manipulate the EXPRESS-defined data in the databases. Depending on the fuzzy object-oriented database platform, various implementation algorithms of these SDAI functions have been developed.

References

Abiteboul, S., Hull, R. and Vianu, V. (1995), Foundations of Databases, Addison-Wesley, Reading, Mass.

Chen, P. P. (1976), The entity-relationship model: toward a unified view of data, ACM Transactions on Database Systems, 1 (1): 9-36.

Eggers, J. (1988), Implementing EXPRESS in SQL, Document TC184/SC4/WG1 N292, ISO.

Goh, A., Hui, S. C., Song, B. and Wang, F. Y. (1994), A study of SDAI implementation on object-oriented databases, Computer Standards & Interfaces, 16: 33-43.

Goh, A., Hui, S. C., Song, B. and Wang, F. Y. (1997), A STEP/EXPRESS to object-oriented databases translator, International Journal of Computer Applications in Technology, 10 (1-2): 90-96.

Hardwick, M. (1991), Implementing the PDES/STEP specifications in an object-oriented database, Proceedings of AUTOFACT'91.

Herbst, A. (1994), Archiving of data in an EXPRESS/SDAI database, Proceedings of EUG '94 – The Fourth EXPRESS Users Group Conference.

IDEF (2000), IDEF Family of Methods, http://www.idef.com/default.html

ISO 10303-203 (1994), Industrial Automation Systems and Integration — Product Data Representation and Exchange — Part 203: Application Protocol: Configuration Controlled Design, ISO, Geneva.

ISO 10303-42 (1994), Industrial Automation Systems and Integration — Product Data Representation and Exchange — Part 42: Integrated Generic Resources: Geometric and Topological Representation, ISO, Geneva.

ISO TC184/SC4 (1998), Industrial Automation Systems and Integration — Product Data Representation and Exchange – Part 26: Interface Definition Language Binding to the Standard Data Access Interface.

ISO TC184/SC4 WG7 N23 (1991), Industrial Automation Systems and Integration – Product Data Representation and Exchange – Functional Requirements for a STEP Data access Interface.

ISO TC184/SC4 WG7 N392 (1995), Industrial Automation Systems and Integration – Product Data Representation and Exchange – Part 22: Implementation Methods: Standard Data Access Interface.

ISO TC184/SC4 WG7 N393 (1995), Industrial Automation Systems and Integration — Product Data Representation and Exchange — Part 23: C++ Language Binding to the Standard Data Access Interface Specification.

ISO TC184/SC4 WG7 N394 (1995), Industrial Automation Systems and Integration – Product Data Representation and Exchange – Part 24: Standard Data Access Interface – C Language Late Binding.

Krebs, T. and Decker, J. (1995), Translating EXPRESS models to the extended relational database management system POSTGRES, Proceedings of EUG '95 —The Fifth EXPRESS Users Group Conference.

Krebs, T. and Lührsen, H. (1995), STEP databases as integration platform for concurrent engineering, Proceedings of 2nd International Conference on Concurrent Engineering, 131-142.

Kusiak, A., Letsche, T. and Zakarian, A. (1997), Data modeling with IDEF1X, International Journal of Computer Integrated Manufacturing, 10: 470-486.

Loffredo, D. (1998), Efficient Database Implementation of EXPRESS Information Models, Ph.D. Thesis, Rensselaer Polytechnic Institute, Troy, New York.

Muller, J. and Smith, G. (1993), A pre-competitive project in intelligent manufacturing technology, Proceedings of AAAI '93 Workshop on Intelligent Manufacturing Technology.

ODMG (2000), Object Data Management Group, http://www.odmg.org/.

Raghavan, V. (1992), STEP Relational Interface, Master's Thesis, Rensselaer Polytechnic Institute, Troy, New York.

Sanderson, D. (1995), Loss of Data Semantics in Syntax Directed Translation, Ph.D. Thesis, Rensselaer Polytechnic Institute, Troy, New York.

Sanderson, D. and Spooner, D. (1993), Mapping between EXPRESS and traditional DBMS models, Proceedings of EUG'93 — The Third EXPRESS Users Group Conference.

Warthen, B. D. (1992), What is STEP for?", Product Data International, 3 (6): 1-10.

Index

1NF (first normal form), 21, 22,
161, 173
α-cut, 51, 52

attribute, 3, 4, 5, 7, 8, 9, 1011, 13,
16, 17, 18, 20, 21, 22, 23, 24,
25, 47, 48, 59, 60, 61, 6263,
67, 68, 70, 79, 80, 81, 82, 87,
89, 93, 94, 95, 102, 103, 104,
105, 106, 107, 108, 109, 117,
126, 127, 130, 132, 134, 138,
139, 140, 141, 142, 144, 145,
146, 147, 148, 149, 150, 151,
152, 153, 154, 155, 160-165,
168, 169, 170, 171, 173, 174,
175, 176, 181, 188, 189, 191,
192, 194, 195, 196, 197, 198,
199, 200, 201, 202
attribute domain, 3, 20, 67, 70,
139, 141, 145, 146, 149, 150,
152, 153, 174, 191
algorithm, 37, 38, 47, 110, 111,
118, 180, 199, 200, 201, 202

CAD/CAPP/CAM, 1
Cartesian product, 20, 21, 22, 140,
145
class
    subclass, 5, 6, 8, 25, 60, 63, 64,
    65, 66, 67, 68, 69, 70, 147,
    148, 149, 151, 152, 153, 154,
    163, 164, 165, 171

superclass, 5, 6, 8, 25, 60, 63,
64, 65, 66, 67, 68, 69, 70, 147,
148, 149, 151, 152, 153, 154,
164, 165, 171, 191
class hierarchy, 147, 190
closeness relation, 141
conceptual design, 26, 30, 34, 35,
36, 59, 77, 133, 134, 135
constraint
    cardinality ratio constraint, 4
    constraint rule, 11, 163, 164,
    165
    integrity constraint, 21, 140,
    162, 191
    participation constraint, 4,
creation, 196, 197, 200, 201, 202

data dependency, 138, 155
data model, 1, 2, 3, 4, 7, 9, 15, 22,
23, 24, 26, 27, 28, 29, 49, 59,
60, 61, 66, 67, 76, 77, 79, 128,
131, 132, 133, 134, 135, 136,
137, 138, 144, 145, 146, 147,
148, 159, 162, 166, 167, 168,
172, 173, 174, 175, 176, 179,
180, 181, 188, 189, 190, 192,
202
databases
    relational databases, 1, 3, 7, 9,
    10, 19-24, 26, 28, 56, 59, 137,
    138, 139, 140, 142, 144, 146,
    155, 159, 160, 161, 162, 166,
    167, 168, 169, 172, 173, 174,